ELECTROMAGNETICS
OF COMPLEX MEDIA
Frequency Shifting by a Transient Magnetoplasma Medium

Dikshitulu K. Kalluri
Department of Electrical Engineering
University of Massachusetts Lowell
Lowell, Massachusetts

CRC Press

Boca Raton Boston London New York Washington, D.C.

Acquiring Editor:	Robert Stern
Project Editor:	Sylvia Wood
Marketing Manager:	Jane Stark
Cover design:	Dawn Boyd
PrePress:	Greg Cuciak

Library of Congress Cataloging-in-Publication Data

Kalluri, Dikshitulu K.
    Electromagnetics of complex media : frequency shifting by a
transient magnetoplasma medium / by Dikshitulu K. Kalluri.
        p.   cm.
    Includes bibliographical references and index.
    ISBN 0-8493-2522-6 (alk. paper)
    1. Electromagnetism.   2. Plasma (Ionized gases)   I. Title.
QC760.K36   1998
537--dc21   98-16443
                                                                          CIP

In Loving Memory

_____

*Of my father, Jagannatham Kalluri and
my mother, Venkatalakshmi Kalluri*

Foreword

I am pleased to write a Foreword to this book by Professor D. K. Kalluri, who has been a friend since I first became aware of his work in 1992. In that year, Professor C. J. Joshi of the University of California Los Angeles and co-workers were undertaking a series of experiments, based on ideas developed by a number of theorists, to study the frequency shift of radiation by moving ionization fronts in plasmas. This work was closely related to that which Professor Kalluri had been doing for many years, and so we arranged for him to spend the summer of 1992 in Berkeley. In this way, a fruitful collaboration and a good friendship were initiated.

The transformation of an electromagnetic wave by an inhomogeneous medium is discussed in many books. The Doppler effect, which is a frequency change due to a moving boundary, is also a standard topic in many books. However, the transformation of the frequency of an electromagnetic wave by a general time-varying medium is rarely discussed. The frequency change in a nonmoving media is contrary to the usual experience of wave change. Because of his background in electrical engineering and electromagnetics, Professor Kalluri could weave, in a unique way, this topic of current research interest into a book on *Electromagnetics of Complex Media* varying in space and time. By using simple ideal models, he has focused on the effect of the time-varying parameters in conjunction with one or more additional kinds of complexities in the properties of the medium, and thus made the subject more accessible.

The reader can be assured that Professor Kalluri is highly competent to write this book, for he has contributed original research papers on the frequency shifting of electromagnetic radiation using a transient magnetized plasma. He has expertise, and many publications, dealing with moving media, the use of Laplace transforms to study the effect of boundaries on transient solution, the generation of a downshifted wave whose frequency can be controlled by the strength of a static magnetic field (transformation of a whistler wave), and the turning-off of the external magnetic field (which showed that an original whistler wave is converted into a wiggler magnetic field). In short, the book that follows is written by a highly competent author who treats important subjects. I hope the reader will enjoy it as much as I have.

Andrew M. Sessler
Berkeley, California

Foreword

Author

Dikshitulu K. Kalluri, Ph.D., is Professor of Electrical and Computer Engineering at the University of Massachusetts Lowell. Born in Chodavaram, India, he received his B.E. degree in electrical engineering from Andhra University, India; a D.I.I Sc. degree in high-voltage engineering from the Indian Institute of Science in Bangalore, India; earned a master's degree in electrical engineering from the University of Wisconsin, Madison, and his doctorate in electrical engineering from the University of Kansas, Lawrence.

Dr. Kalluri began his career at the Birla Institute, Ranchi, India, advancing to the rank of Professor, heading the Electrical Engineering Department, then serving as (Dean) Assistant Director of the institute. He has collaborated with research groups at the Lawrence Berkeley Laboratory, the University of California Los Angeles, the University of Southern California, and the University of Tennessee, and has worked several summers as a faculty research associate at Air Force Laboratories. Since 1984, he has been with the University of Massachusetts Lowell, where he is coordinator of the doctoral program and co-director of the Center for Electromagnetic Materials and Optical Systems (CEMOS). As part of the center, he recently established the Electromagnetics and Complex Media Research Laboratory.

Dr. Kalluri, a fellow of the Institute of Electronic and Telecommunication Engineers and a member of Eta Kappa Nu and Sigma Xi, has published many technical articles and reviews.

Preface

After careful thought I have developed two graduate courses in electromagnetics, each with a different focus. One of them assumes a simple electromagnetic medium, a medium described by scalar permittivity (ε), permeability (μ), and conductivity (σ). The geometrical effect, due to the shape and the size of the boundaries, is the main focus. There are many excellent textbooks, for example, *Advanced Electromagnetics* by Balanis, to serve the needs of this course.

Recent advances in material science suggest that materials can be synthesized with any desired electromagnetic properties. The optimum properties for a given application must be understood and sought. To facilitate developing such an appreciation in our graduate students, I have developed another course called the "Electromagnetics of Complex Media." The focus here is to bring out the major effects due to each kind of complexity in the medium properties. The medium is considered complex if any of the electromagnetic parameters ε, μ, and σ are not scalar constants. If the parameters are functions of signal frequency, we have temporal dispersion. If the parameters are tensors, we have anisotropy. If the parameters are functions of position, we have inhomogeneity. A plasma column in the presence of a static magnetic field is at once dispersive, anisotropic, and inhomogeneous. For this reason, I have chosen plasma as the basic medium to illustrate some aspects of the transformation of an electromagnetic wave by a complex medium.

An additional aspect of medium complexity that is of current research interest arises out of the time-varying parameters of the medium. Powerful lasers that produce ultrashort pulses for ionizing gases into plasmas can permit fast changes in the dielectric constant of a medium. An idealization of this process is called the sudden creation of the plasma or sudden switching of the medium. More practical processes require a model of a time-varying plasma with an *arbitrary rise time*.

The early chapters of this book use a mathematical model that usually has one kind of complexity. The medium is often assumed to be unbounded in space or has a simple plane boundary. The field variables and the parameters are often assumed to vary in one spatial coordinate. This eliminates the use of heavy mathematics and permits the focus to be on the effect. The last chapter, however, has a section on the use of the finite-difference time-domain method for the numerical simulation of three-dimensional problems.

The main effect of switching a medium is to shift the frequency of the source wave. The frequency change is contrary to the usual experience of wave change we find. The exception is the Doppler effect, which is a frequency change caused by a moving boundary. The moving boundary is a particular case of a time-varying medium.

The primary title indicates that this book will serve the needs of students who study electromagnetics as a basis for a number of disciplines that use "complex materials." Examples are electro-optics, plasma science and engineering, microwave engineering, and solid-state devices. The aspects of electromagnetic wave transformation by a complex medium that are emphasized in the book are

1. dispersive medium
2. tunneling of power through a plasma slab by evanescent waves
3. characteristic waves in an anisotropic medium
4. transient medium and frequency shifting
5. Green's function for unlike anisotropic media
6. perturbation technique for unlike anisotropic media
7. adiabatic analysis for modified source wave

All the above topics use one-dimensional models.

The following topics are covered briefly in this book: (1) chiral media, (2) surface waves, and (3) periodic media. The topics that are not covered include (1) nonlinear media, (2) parametric instabilities, and (3) random media. I hope to include these topics in the future in an expanded version of the book to serve a two-semester course or to give a choice of topics for a one-semester course.

Problems are added at the end of the book for the benefit of those who would like to use the book as a textbook. The background needed is a one-semester undergraduate electromagnetics course that includes a discussion of plane waves in a simple medium. With this background, a senior undergraduate student or a first-year graduate student can easily follow the book. The solution manual for the problems is available.

The secondary title of the book emphasizes the viewpoint of frequency change and is intended to draw the attention of new researchers who wish to have a quick primer on the theory of using magnetoplasmas for coherent generation of tunable radiation. I hope the book will stimulate experimental and additional theoretical and numerical work on the remarkable effects that can be obtained by the temporal and spatial modification of the magnetoplasma parameters. A large part of the book contains research published by a number of people, including the author of this book, in recent issues of several research journals. Particular attention is drawn to the reprints given in Appendixes B through H. The book also contains a number of unpublished results.

The Rationalized MKS system of units is used throughout the book. The harmonic time and space variations are denoted by $\exp(j\omega t)$ and $\exp(-jkz)$, respectively.

Acknowledgments

I am proud of my present and past doctoral students who have shared with me the trials and tribulations of exploring a new area of research. Among them, I particularly would like to mention Drs. V. R. Goteti and T. T. Huang, and Mr. Joo Hwa Lee. A very special thanks to Joo Hwa, who, in his last semester of doctoral work, undertook the task of critically reviewing the manuscript and used his considerable computer skills to format the text and the figures to the specified standard. I am grateful to the University of Massachusetts Lowell for granting a sabbatical during the spring of 1996 for the purpose of writing this book. The support of my research by the Air Force Office of Scientific Research and the Air Force Laboratories during 1996–1997 is gratefully acknowledged. I am particularly thankful to Dr. K. M. Groves for acting as my focal point at the laboratory and for contributing to the research. The encouragement of my friends and collaborators, Professor Andrew Sessler of the University of California, and Professor Igor Alexeff of the University of Tennessee, helped me a great deal in doing my research and in writing this book.

My peers with whom I had opportunity to discuss research of mutual interest include Professors A. Baños, Jr., S. A. Bowhill, J. M. Dawson, M. A. Fiddy, O. Ishihara, C. J. Joshi, T. C. Katsoules, H. H. Kuehl, S. P. Kuo, M. C. Lee, W. B. Mori, A. G. Nerukh, E. J. Powers, Jr., T. C. K. Rao, B. Reinisch, G. Sales, B. V. Stanic, N. S. Stepanov, D. Wunsch, and B. J. Wurtele and Drs. V. W. Byszewski, S. J. Gitomer, P. Muggli, R. L. Savage, Jr., and S. C. Wilks. The anonymous reviewers of our papers also belong to this group. This group played an important role in providing the motivation to continue our research.

My special thanks go to Dr. Robert Stern at CRC Press for waiting for the manuscript and, when once submitted, processing it with great speed. I appreciate the help and advice given by project editor Sylvia Wood, also at CRC.

Finally, I am most thankful to my wife, Kamala, for assisting me with many aspects of writing the book and to my children Srinath, Sridhar, and Radha for standing by me, encouraging me, and giving up their share of my time for the sake of research.

Dikshitulu K. Kalluri

Contents

1   Isotropic Plasma: Dispersive Medium

1.1   INTRODUCTION

Plasma is a *quasineutral* mixture of charged particles and neutral particles. It is characterized by two independent parameters for each of the particle species: the particle density N and the temperature T. Plasma physics deals with such mixtures, and there is a vast amount of literature on this topic. A few references of direct interest to the reader of this book are given at the end of this chapter. References 1–3 deal with modeling of a magnetized plasma as an electromagnetic medium. The models are adequate in exploring some of the applications in which the medium can be considered to have time-invariant electromagnetic parameters.

There are some applications for which the thermal effects are unimportant; such a plasma is called a *cold plasma*. A *Lorentz plasma*[1] is a further simplification of the medium. In this model it is assumed that the electrons interact with each other only through collective space charge forces and that the heavy positive ions and neutral particles are at rest. The positive ions serve as a background that ensures the overall charge neutrality of the mixture. In this book the Lorentz plasma will be the dominant model used to explore the major effects of a nonperiodically time-varying electron density profile $N(t)$. Departure from the model will be made only when necessary to bring in other relevant effects. References 1 and 2 show the approximations made to arrive at the Lorentz plasma model.

1.2   BASIC FIELD EQUATIONS FOR A COLD ISOTROPIC PLASMA

The electric field $\mathbf{E}(\mathbf{r},t)$, the magnetic field $\mathbf{H}(\mathbf{r},t)$, and the velocity field $\mathbf{v}(\mathbf{r},t)$ of the electrons in the isotropic Lorentz plasma satisfy the following equations:

$$\nabla \times \mathbf{E} = -\mu_0 \frac{\partial \mathbf{H}}{\partial t}, \tag{1.1}$$

$$\nabla \times \mathbf{H} = \varepsilon_0 \frac{\partial \mathbf{E}}{\partial t} + \mathbf{J}, \tag{1.2}$$

$$m \frac{d\mathbf{v}}{dt} = -q\mathbf{E}, \tag{1.3}$$

where $\mathbf{J}$ is the free-electron current density in the plasma, q is the absolute value of the charge of an electron, and m is the mass of an electron. The relation between

1

the current density and the electric field in the plasma will depend on the ionization process[4-7] that creates the plasma. Assume that the electron density profile $N(t)$ in the plasma is known and that the created electrons have zero velocity at the instant of their birth, then from Equation 1.3 the velocity at t of the electrons born at t_i is given by

$$\mathbf{v}_i(\mathbf{r},t) = -\frac{q}{m}\int_{t_i}^t \mathbf{E}(\mathbf{r},\tau)d\tau. \tag{1.4}$$

The change in current density at t due to the electrons born at t_i can be computed as follows:

$$\Delta\mathbf{J}(\mathbf{r},t) = -q\Delta N_i \mathbf{v}_i(\mathbf{r},t). \tag{1.5}$$

Here ΔN_i is the electron density added at t_i and is given by

$$\Delta N_i = \left[\frac{\partial N}{\partial t}\right]_{t=t_i} \Delta t_i. \tag{1.6}$$

Therefore, the current density is given by (see Appendix A)

$$\mathbf{J}(\mathbf{r},t) = \frac{q^2}{m}\int_0^t \frac{\partial N(\mathbf{r},\tau)}{\partial\tau}\,d\tau\int_\tau^t \mathbf{E}(\mathbf{r},\alpha)d\alpha + \mathbf{J}(\mathbf{r},0). \tag{1.7}$$

The expression for $\mathbf{J}$ may be simplified (see Appendix A):

$$\mathbf{J}(\mathbf{r},t) = \varepsilon_0\int_0^t \omega_p^2(\mathbf{r},\tau)\mathbf{E}(\mathbf{r},\tau)d\tau + \mathbf{J}(\mathbf{r},0). \tag{1.8}$$

Here, ω_p^2 is the square of the plasma frequency proportional to the electron density N and is given by

$$\omega_p^2(\mathbf{r},t) = \frac{q^2 N(\mathbf{r},t)}{m\varepsilon_0}. \tag{1.9}$$

See Reference 1 for a physical explanation of the term *plasma frequency*. By differentiating Equation 1.8 a differential equation for $\mathbf{J}$ is obtained:

$$\frac{d\mathbf{J}}{dt} = \varepsilon_0\omega_p^2(\mathbf{r},t)\mathbf{E}(\mathbf{r},t). \tag{1.10}$$

Thus, the equations in a time-varying and space-varying isotropic plasma are Equations 1.1, 1.2, and 1.10. This description used the variable $\mathbf{J}$ instead of $\mathbf{v}$. By taking the curl of Equation 1.1 and eliminating $\mathbf{H}$ and $\mathbf{J}$ by using Equations 1.2 and 1.10, the wave equation for $\mathbf{E}$ can be derived:

$$\nabla^2\mathbf{E} - \nabla(\nabla\cdot\mathbf{E}) - \frac{1}{c^2}\frac{\partial^2\mathbf{E}}{\partial t^2} - \frac{1}{c^2}\omega_p^2(\mathbf{r},t)\mathbf{E} = 0. \tag{1.11}$$

Similar efforts will lead to a wave equation for the magnetic field:

$$\nabla^2\dot{\mathbf{H}} - \frac{1}{c^2}\frac{\partial^2\dot{\mathbf{H}}}{\partial t^2} - \frac{1}{c^2}\omega_p^2(\mathbf{r},t)\dot{\mathbf{H}} + \varepsilon_0\nabla\omega_p^2(\mathbf{r},t)\times\mathbf{E} = 0, \tag{1.12}$$

where

$$\dot{\mathbf{H}} = \frac{\partial\mathbf{H}}{\partial t}. \tag{1.13}$$

In deriving Equation 1.12, the equation

$$\nabla\cdot\mathbf{H} = 0 \tag{1.14}$$

is used. The last term in Equation 1.12 contains $\mathbf{E}$; however, if ω_p^2 varies only with t, its gradient is zero and Equation 1.12 becomes

$$\nabla^2\dot{\mathbf{H}} - \frac{1}{c^2}\frac{\partial^2\dot{\mathbf{H}}}{\partial t^2} - \frac{1}{c^2}\omega_p^2(t)\dot{\mathbf{H}} = 0. \tag{1.15}$$

1.3 ONE-DIMENSIONAL EQUATIONS

Consider the particular case where (a) the variables are functions of one spatial coordinate only, say, the z coordinate, (b) the electric field is linearly polarized in the x-direction, and (c) the variables are denoted by

$$\mathbf{E} = \hat{x}E(z,t), \tag{1.16}$$

$$\mathbf{H} = \hat{y}H(z,t), \tag{1.17}$$

$$\mathbf{J} = \hat{x}J(z,t), \tag{1.18}$$

$$\omega_p^2 = \omega_p^2(z,t). \tag{1.19}$$

The basic equations for E, H, and J take the following simple form:

$$\frac{\partial E}{\partial z} = -\mu_o \frac{\partial H}{\partial t}, \tag{1.20}$$

$$-\frac{\partial H}{\partial z} = \varepsilon_0 \frac{\partial E}{\partial t} + J, \tag{1.21}$$

$$\frac{dJ}{dt} = \varepsilon_0 \omega_p^2(z,t)E, \tag{1.22}$$

$$\frac{\partial^2 E}{\partial z^2} - \frac{1}{c^2}\frac{\partial^2 E}{\partial t^2} - \frac{1}{c^2}\omega_p^2(z,t)E = 0, \tag{1.23}$$

$$\frac{\partial^2 \dot{H}}{\partial z^2} - \frac{1}{c^2}\frac{\partial^2 \dot{H}}{\partial t^2} - \frac{1}{c^2}\omega_p^2(z,t)\dot{H} + \varepsilon_0\frac{\partial}{\partial z}\omega_p^2(z,t)E = 0. \tag{1.24}$$

If ω_p^2 varies only with t, Equation 1.24 reduces to

$$\frac{\partial^2 \dot{H}}{\partial z^2} - \frac{1}{c^2}\frac{\partial^2 \dot{H}}{\partial t^2} - \frac{1}{c^2}\omega_p^2(t)\dot{H} = 0. \tag{1.25}$$

Next consider that the electric field is linearly polarized in the y-direction, i.e.,

$$\mathbf{E} = \hat{y}E(z,t), \tag{1.26}$$

$$\mathbf{H} = -\hat{x}H(z,t), \tag{1.27}$$

$$\mathbf{J} = \hat{y}J(z,t). \tag{1.28}$$

The basic equations for E, H, J are given by Equations 1.20 through 1.25. It immediately follows that the basic equations for circular polarization, where

$$\mathbf{E} = \left(\hat{x} \mp j\hat{y}\right)E(z,t), \tag{1.29}$$

$$\mathbf{H} = \left(\pm\hat{x} + j\hat{y}\right)H(z,t), \tag{1.30}$$

$$\mathbf{J} = \left(\hat{x} \mp j\hat{y}\right)J(z,t), \tag{1.31}$$

are once again given by Equations 1.20 through 1.25. In the above, the upper sign is for the right circular polarization and the lower sign for the left circular polarization.

There is one more one-dimensional solution that has some physical significance. Let

$$\mathbf{E} = \hat{z}E(z,t),$$

(1.32)

$$\mathbf{H} = 0,$$

(1.33)

$$\mathbf{J} = \hat{z}J(z,t).$$

(1.34)

From Equation 1.32, it follows that

$$\nabla \times \mathbf{E} = 0,$$

(1.35)

and from Equations 1.2, 1.3, and 1.10

$$\varepsilon_0 \frac{\partial E}{\partial t} = -J,$$

(1.36)

$$m \frac{d\mathbf{v}}{dt} = -q\mathbf{E},$$

(1.37)

$$\frac{dJ}{dt} = \varepsilon_0 \omega_p^2(z,t)E.$$

(1.38)

Also,

$$\nabla \cdot \mathbf{E} = \frac{\partial E}{\partial z}.$$

(1.39)

From Equation 1.11,

$$\frac{\partial^2 E}{\partial t^2} + \omega_p^2(z,t)E = 0.$$

(1.40)

Equation 1.40 can also be obtained from Equations 1.37 and 1.38. Equation 1.40 can be easily solved if ω_p^2 is not a function of time:

$$E = E_0(z)\cos\left[\omega_p(z)t\right],$$

(1.41)

$$\mathbf{E} = \hat{z}E.$$

(1.42)

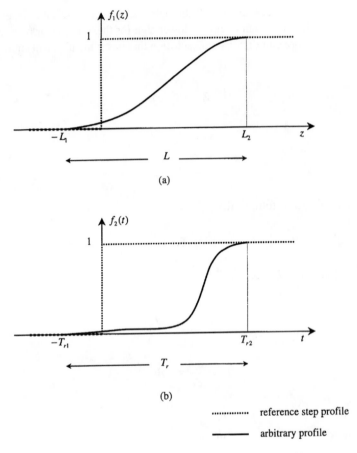

FIGURE 1.1   (a) Spatial profile. (b) Temporal profile.

Equation 1.42 represents a longitudinal oscillation of the electrostatic field at the plasma frequency. In a warm plasma, this oscillation will be converted into an electrostatic wave, which is also called the *Langmuir wave* or the electron plasma wave.[1] For a warm plasma, Equation 1.3 needs modification because of the pressure gradient forces that are caused by the thermal effects. In metals at optical frequencies, the *Fermi velocity* plays the role of a thermal velocity. See Reference 8.

1.4   PROFILE APPROXIMATIONS FOR SIMPLE SOLUTIONS

The subsequent chapters discuss solutions based on some simple approximations for the ω_p^2 profile as it transitions from one medium to the other. Asymptotic values for it are assumed deep inside each medium and a nonperiodic change of the parameter from medium 1 to medium 2 over a length scale L (range length) and a timescale T_r (rise time). Figure 1.1 and Equations 1.43 through 1.47 describe the profiles:

$$\omega_p^2(z,t) = \omega_{p1}^2 + \left(\omega_{p2}^2 - \omega_{p1}^2\right) f_1(z) f_2(t), \tag{1.43}$$

$$f_1(z) = 0, \quad -\infty < z < -L_1, \tag{1.44}$$

$$= 1, \quad L_2 < z < \infty, \tag{1.45}$$

$$f_2(t) = 0, \quad -\infty < t < -T_{r1}, \tag{1.46}$$

$$= 1, \quad T_{r2} < t < \infty. \tag{1.47}$$

In the above, ω_{p1}^2 is the plasma frequency of the medium 1 as $z \to -\infty$ and $t \to -\infty$ and ω_p^2 is the plasma frequency of the medium 2 as $z \to \infty$ and $t \to \infty$. The functions $f_1(z)$ and $f_2(t)$, sketched in Figure 1.1, describe the transition of the plasma frequency from its asymptotic values.

If the range length $L_1 + L_2 = L$ is much less than the significant length of the problem, say, the wavelength λ of the source wave in medium 1, then the profile can be approximated as a spatial step profile and the problem solved as a boundary value problem using appropriate boundary conditions. If $L \gg \lambda$, then adiabatic analysis based on the approximation techniques like Wentzel–Kramers–Brillouin (WKB) can be used. If L is comparable to λ, a perturbation technique[9] based on Green's function for a sharp boundary of unlike media can be used. Reference 9 gives an excellent account of the application of this technique to the calculation of the reflection coefficient for a spatial dielectric profile. The inhomogeneous media problem (spatial profile) was investigated extensively in view of its applications in optics[9] and ionospheric physics.[10]

If the rise time $T_{r1} + T_{r2} = T_r$ is much less than the significant period of the problem, say, the period t_0 of the source wave in medium 1, then the profile can be approximated as a temporal step profile and the problem solved as an initial value problem using appropriate initial conditions. If $T_r \gg t_0$, then adiabatic analysis based on approximation techniques like WKB can be used. If the rise time T_r is comparable to t_0, a perturbation technique based on Green's function for the switched medium can be used. These are some of the techniques explored in subsequent chapters.

Another simple profile of great interest is the ionization front profile (Figure 1.2), defined by

$$\omega_p^2(z,t) = \omega_{p0}^2 \left[u\left(z + v_F t\right)\right]. \tag{1.48}$$

In the above u is a unit step function

$$u(\varphi) = 0, \quad \varphi < 0,$$
$$= 1, \quad \varphi \geq 0, \tag{1.49}$$

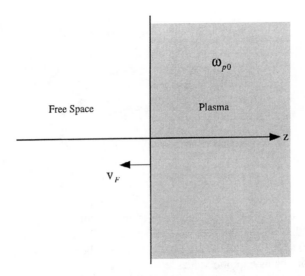

FIGURE 1.2   Ionization front.

and v_F is the velocity of the ionization front. Such a profile can be created by a source of ionizing radiation pulse, say, a strong laser pulse. As the pulse travels in a neutral gas, it converts it into a plasma, thus creating a moving boundary between the plasma and the unionized gas. Section 10.3 deals with such a problem. In passing, it may be mentioned that the ionization front problem is somewhat different from the moving plasma problem. In the front problem the boundary alone is moving and the plasma is not moving with the boundary. The problem of the moving plasma is dealt with by a number of authors[11-15] and is not discussed in this book.

1.5   DISPERSIVE MEDIA

The dielectric constant ε_r of a dielectric is frequency independent over wide bands of frequencies, and in these bands the dielectric is nondispersive. In this book the term "pure dielectric" or simply "dielectric" is used to describe a nondispersive dielectric with a dielectric constant greater than or equal to 1.

Plasma can be modeled as a dielectric. However, its dielectric constant depends on the frequency of the signal and is given by

$$\varepsilon_p = \left(1 - \frac{\omega_p^2}{\omega^2}\right). \tag{1.50}$$

This dependence makes it highly dispersive and leads to a qualitative difference in the properties of the waves scattered by a temporal change in the electron density of the plasma medium as compared with such scattering by a pure dielectric.

In the presence of a static magnetic field the dielectric constant is not only a function of frequency; it also becomes a tensor and the medium has to be modeled as a dispersive anisotropic medium. The properties of the waves scattered by a temporal change in the electron density of the plasma medium in the presence of a static magnetic field involve many more modes.[16] The subsequent chapters deal with these topics.

REFERENCES

1. Heald, M. A. and Wharton, C. B., *Plasma Diagnostics with Microwaves*, Wiley, New York, 1965.
2. Booker, H. G., *Cold Plasma Waves*, Kluwer, Hingham, MA, 1984.
3. Tannenbaum, B. S., *Plasma Physics*, McGraw-Hill, New York, 1967.
4. Baños, A., Jr., Mori, W. B., and Dawson, J. M., Computation of the electric and magnetic fields induced in a plasma created by ionization lasting a finite interval of time, *IEEE Trans. Plasma Sci.*, 21, 57, 1993.
5. Kalluri, D. K., Goteti, V. R., and Sessler, A. M., WKB solution for wave propagation in a time-varying magnetoplasma medium: longitudinal propagation, *IEEE Trans. Plasma Sci.*, 21, 70, 1993.
6. Lampe, M. and Ott, E., Interaction of electromagnetic waves with a moving ionization front, *Phys. Fluids*, 21, 42, 1978.
7. Stepanov, N. S, Dielectric constant of unsteady plasma, *Sov. Radiophys. Quant. Electr.*, 19, 683, 1976.
8. Forstmann, F. and Gerhardts, R. R., *Metal Optics Near the Plasma Frequency*, Springer-Verlag, New York, 1986.
9. Lekner, J., *Theory of Reflection*, Kluwer, Boston, 1987.
10. Budden, K. G., *Radio Wave in the Ionosphere*, Cambridge University Press, Cambridge, 1961.
11. Chawla, B. R. and Unz, H., *Electromagnetic Waves in Moving Magneto-plasmas*, The University Press of Kansas, Lawrence, 1971.
12. Kalluri, D. K. and Shrivastava, R. K., Radiation pressure due to plane electromagnetic waves obliquely incident on moving media, *J. Appl. Phys.*, 49, 3584, 1978.
13. Kalluri, D. K. and Shrivastava, R. K., On total reflection of electromagnetic waves from moving plasmas, *J. Appl. Phys.*, 49, 6169, 1978.
14. Kalluri, D. K. and Shrivastava, R. K., Brewster angle for a plasma medium moving at relativistic speed, *J. Appl. Phys.*, 46, 1408, 1975.
15. Kalluri, D. K. and Shrivastava, R. K., Electromagnetic wave interaction with moving bounded plasmas, *J. Appl. Phys.*, 44, 4518, 1973.
16. Kalluri, D. K., Frequency-shifting using magnetoplasma medium, *IEEE Trans. Plasma Sci.*, 21, 77, 1993.

2 Space-Varying and Time-Invariant Isotropic Medium

2.1 BASIC EQUATIONS

If it is assumed that the electron density varies only in space (inhomogeneous isotropic plasma), then from Equation 1.9

$$\omega_p^2 = \omega_p^2(\mathbf{r}).$$ (2.1)

Basic solutions can then be constructed by assuming that the field variables vary harmonically in time:

$$\mathbf{F}(\mathbf{r},t) = \mathbf{F}(\mathbf{r})\exp(j\omega t).$$ (2.2)

In the above, $\mathbf{F}$ stands for any of $\mathbf{E}$, $\mathbf{H}$, or $\mathbf{J}$. Equations 1.1, 1.2, and 1.10 then reduce to Equations 2.3 through 2.5:

$$\nabla \times \mathbf{E} = -j\omega\mu_0\mathbf{H},$$ (2.3)

$$\nabla \times \mathbf{H} = j\omega\varepsilon_0\mathbf{E} + \mathbf{J},$$ (2.4)

$$j\omega\mathbf{J} = \varepsilon_0\omega_p^2(\mathbf{r})\mathbf{E}.$$ (2.5)

By combining Equations 2.4 and 2.5, the following can be written:

$$\nabla \times \mathbf{H} = j\omega\varepsilon_0\varepsilon_p(\mathbf{r},\omega)\mathbf{E},$$ (2.6)

where ε_p is the dielectric constant of the isotropic plasma and is given by

$$\varepsilon_p(\mathbf{r},\omega) = 1 - \frac{\omega_p^2(\mathbf{r})}{\omega^2}.$$ (2.7)

The wave equations 1.11 and 1.12 reduce to

$$\nabla^2 \mathbf{E} + \frac{\omega^2}{c^2}\varepsilon_p(\mathbf{r})\mathbf{E} = \nabla(\nabla \cdot \mathbf{E}), \tag{2.8}$$

$$\nabla^2 \mathbf{H} + \frac{\omega^2}{c^2}\varepsilon_p(\mathbf{r})\mathbf{H} = -\frac{\varepsilon_o}{j\omega}\nabla\omega_p^2(\mathbf{r})\times\mathbf{E} = -j\omega\varepsilon_0\nabla\varepsilon_p(\mathbf{r})\times\mathbf{E}. \tag{2.9}$$

The one-dimensional equations 1.20 through 1.22 take the form:

$$\frac{\partial E}{\partial z} = -j\omega\mu_0 H, \tag{2.10}$$

$$-\frac{\partial H}{\partial z} = j\omega\varepsilon_0 E + J, \tag{2.11}$$

$$j\omega J = \varepsilon_0 \omega_p^2(z)E. \tag{2.12}$$

By combining Equations 2.11 and 2.12 or from Equation 2.6,

$$-\frac{\partial H}{\partial z} = j\omega\varepsilon_0\varepsilon_p(z,\omega)E, \tag{2.13}$$

where

$$\varepsilon_p(z,\omega) = 1 - \frac{\omega_p^2(z)}{\omega^2}. \tag{2.14}$$

Figure 2.1 shows the variation of ε_p with ω. ε_p is negative for $\omega < \omega_p$, 0 for $\omega = \omega_p$, and positive but less than 1 for $\omega > \omega_p$. The one-dimensional wave equations 1.23 and 1.24, in this case, reduce to Equations 2.15 and 2.16:

$$\frac{d^2 E}{dz^2} + \frac{\omega^2}{c^2}\varepsilon_p(z,\omega)E = 0, \tag{2.15}$$

$$\frac{d^2 H}{dz^2} + \frac{\omega^2}{c^2}\varepsilon_p(z,\omega)H = \frac{1}{\varepsilon_p(z,\omega)}\frac{d\varepsilon_p(z,\omega)}{dz}\frac{dH}{dz}. \tag{2.16}$$

Equation 2.15 can be easily solved if it is further assumed that the dielectric is homogeneous (ε_p is not a function of z). Such a solution is given by

$$E = E^+(z) + E^-(z), \tag{2.17}$$

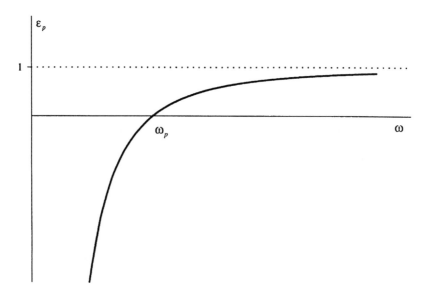

FIGURE 2.1 Dielectric constant vs. frequency for a plasma medium.

$$E = E_0^+ \exp(-jkz) + E_0^- \exp(+jkz), \tag{2.18}$$

where

$$k = \frac{\omega}{c}\sqrt{\varepsilon_p} = k_0\sqrt{\varepsilon_p} = k_0 n. \tag{2.19}$$

Here, n is the refractive index and k_0 is the free-space wave number. The first term on the right side of Equation 2.17 represents a traveling wave in a positive z-direction (positive-going wave) of angular frequency ω and wave number k, and the second term a similar but negative-going wave. The z-direction is the direction of phase propagation and the velocity of phase propagation v_p (phase velocity) of either of the waves is given by

$$v_p = \frac{\omega}{k}. \tag{2.20}$$

From Equation 2.10

$$H^+(z) = \frac{1}{(-j\omega\mu_0)}(-jk)E_0^+ \exp(-jkz) = \frac{1}{\eta}E^+(z), \tag{2.21}$$

$$H^-(z) = -\frac{1}{\eta}E^-(z), \tag{2.22}$$

where

$$\eta = \sqrt{\frac{\mu_0}{\varepsilon_0 \varepsilon_p}} = \frac{\eta_0}{n} = \frac{120\pi}{n} \text{ (ohms)}. \qquad (2.23)$$

Here, η is the intrinsic impedance of the medium. The waves described above, which are one-dimensional solutions of the wave equation in an unbounded isotropic homogeneous medium, are called *uniform plane waves* in the sense that the phase and the amplitude of the waves are constant in a plane. Such one-dimensional solutions in a coordinate-free description are called transverse electric and magnetic (TEM) waves, and for an arbitrarily directed wave their properties may be summarized as follows:

$$\hat{\mathbf{E}} \times \hat{\mathbf{H}} = \hat{k}, \qquad (2.24)$$

$$\hat{\mathbf{E}} \cdot \hat{k} = 0, \ \hat{\mathbf{H}} \cdot \hat{k} = 0, \qquad (2.25)$$

$$E = \eta H. \qquad (2.26)$$

In the above, $\hat{k}$ is a unit vector in the direction of phase propagation. Stated in words, the properties are:

1. Unit electric field vector, unit magnetic field vector, and the unit vector in the direction of (phase) propagation form a mutually orthogonal system.
2. There is no component of the electric or the magnetic field vector in the direction of propagation.
3. The ratio of the electric field amplitude to the magnetic field amplitude is given by the intrinsic impedance of the medium.

Figure 2.2 shows the variation of the dielectric constant ε_p with frequency for a typical real dielectric. For such a medium, in a broad frequency band, ε_p can be treated as not varying with ω and can be denoted by the dielectric constant ε_r. The step profile approximation for a dielectric profile will be considered next. In each of the media 1 and 2, the dielectric may be treated as homogeneous. The solution for this problem is well known but is included here to provide a comparison with the solution for a temporal step profile discussed in Chapter 3.

2.2   DIELECTRIC-DIELECTRIC SPATIAL BOUNDARY

The geometry of the problem is shown in Figure 2.3. Let the incident wave in medium 1, also called the source wave, have the fields:

$$\mathbf{E}_i(x, y, z, t) = \hat{x} E_0 \exp\left[j(\omega_i t - k_i z) \right], \qquad (2.27)$$

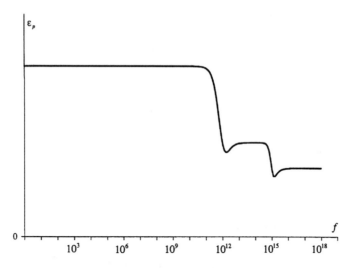

FIGURE 2.2 Sketch of the dielectric constant of a typical material.

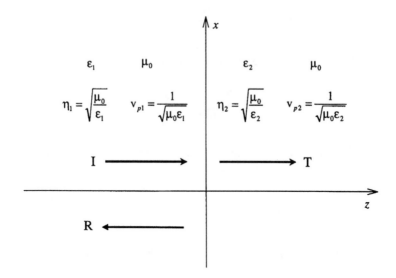

FIGURE 2.3 Dielectric–dielectric spatial-boundary problem.

$$H_i(x,y,z,t) = \hat{y}\frac{E_0}{\eta_1}\exp\left[j(\omega_i t - k_i z)\right],$$ (2.28)

where ω_i is the frequency of the incident wave and

$$k_i = \omega_i\sqrt{\mu_0\varepsilon_1} = \frac{\omega_i}{v_{p1}}.$$ (2.29)

The fields of the reflected wave are given by

$$\mathbf{E}_r(x,y,z,t) = \hat{x} E_r \exp\left[j\left(\omega_r t + k_r z\right)\right],$$ (2.30)

$$\mathbf{H}_r(x,y,z,t) = -\hat{y}\frac{E_r}{\eta_1} \exp\left[j\left(\omega_r t + k_r z\right)\right],$$ (2.31)

where ω_r is the frequency of the reflected wave and

$$k_r = \omega_r \sqrt{\mu_0 \varepsilon_1}.$$ (2.32)

The fields of the transmitted wave are given by

$$\mathbf{E}_t(x,y,z,t) = \hat{x} E_t \exp\left[j\left(\omega_t t - k_t z\right)\right],$$ (2.33)

$$\mathbf{H}_t(x,y,z,t) = \hat{y}\frac{E_t}{\eta_2} \exp\left[j\left(\omega_t t - k_t z\right)\right],$$ (2.34)

where ω_t is the frequency of the transmitted wave and

$$k_t = \omega_t \sqrt{\mu_0 \varepsilon_2} = \frac{\omega_t}{v_{p2}}.$$ (2.35)

The boundary condition of the continuity of the tangential component of the electric field at the interface $z = 0$ can be stated as

$$\mathbf{E}(x,y,0^-,t) \times \hat{z} = \mathbf{E}(x,y,0^+,t) \times \hat{z},$$ (2.36)

$$\left[\mathbf{E}_i(x,y,0^-,t) + \mathbf{E}_r(x,y,0^-,t)\right] \times \hat{z} = \left[\mathbf{E}_t(x,y,0^+,t)\right] \times \hat{z}.$$ (2.37)

The above must be true for all x, y, and t. Thus,

$$E_0 \exp\left[j\omega_i t\right] + E_r \exp\left[j\omega_r t\right] = E_t \exp\left[j\omega_t t\right].$$ (2.38)

Since Equation 2.38 must be satisfied for all t, the coefficients of t in the exponents of Equation 2.38 must match:

$$\omega_i = \omega_r = \omega_t = \omega.$$ (2.39)

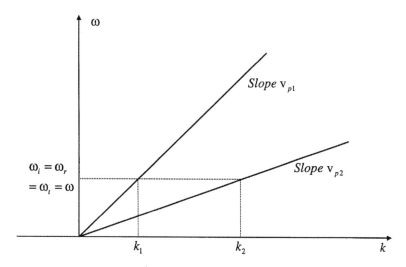

FIGURE 2.4 Conservation of frequency across a spatial boundary.

The above result can be stated as follows: the frequency ω is conserved across a spatial discontinuity in the properties of an electromagnetic medium. As the wave crosses from one medium to the other in space, the wave number k changes as dictated by the change in the phase velocity. See Figure 2.4.

From Equations 2.39 and 2.36,

$$E_0 + E_r = E_t. \qquad (2.40)$$

The second independent boundary condition of continuity of the tangential magnetic field component is written as Equation 2.41:

$$\left[\mathbf{H}_i(x,y,0^-,t)+\mathbf{H}_r(x,y,0^-,t)\right]\times\hat{z}=\left[\mathbf{H}_t(x,y,0^+,t)\right]\times\hat{z}. \qquad (2.41)$$

The reflection coefficient $R_A = E_r/E_0$ and the transmission coefficient $T_A = E_t/E_0$ are determined using Equations 2.40 and 2.41. From Equation 2.41,

$$\frac{E_0 - E_r}{\eta_1} = \frac{E_t}{\eta_2}. \qquad (2.42)$$

The results are

$$R_A = \frac{\eta_2 - \eta_1}{\eta_2 + \eta_1} = \frac{n_1 - n_2}{n_1 + n_2}, \qquad (2.43)$$

$$T_A = \frac{2\eta_2}{\eta_2 + \eta_1} = \frac{2n_1}{n_1 + n_2}. \qquad (2.44)$$

The significance of the subscript A in the above is to distinguish the coefficients from the reflection and transmission coefficients due to a temporal discontinuity in the properties of the medium. This aspect is discussed in Chapter 3.

Next is shown that the time-averaged power density of the source wave is equal to the sum of the time-averaged power density of the reflected wave and the transmitted wave.

$$\left|\tfrac{1}{2}\operatorname{Re}\left[\mathbf{E}_i\times\mathbf{H}_i^*\cdot\hat{z}\right]\right|=\left|\tfrac{1}{2}\operatorname{Re}\left[\mathbf{E}_r\times\mathbf{H}_r^*\cdot\hat{z}\right]\right|+\left|\tfrac{1}{2}\operatorname{Re}\left[\mathbf{E}_t\times\mathbf{H}_t^*\cdot\hat{z}\right]\right|. \tag{2.45}$$

The left side is

$$\text{LHS}=\frac{1}{2}E_0H_0=\frac{1}{2}\frac{E_0^2}{\eta_1}, \tag{2.46}$$

whereas the right side is

$$\text{RHS}=\frac{1}{2}E_0^2\left[\frac{\left|R_A\right|^2}{\eta_1}+\frac{\left|T_A\right|^2}{\eta_2}\right]=\frac{1}{2}\frac{E_0^2}{\eta_1}\left[\frac{\left(\eta_2-\eta_1\right)^2+4\eta_1\eta_2}{\left(\eta_2+\eta_1\right)^2}\right]=\frac{1}{2}\frac{E_0^2}{\eta_1}. \tag{2.47}$$

2.3   REFLECTION BY A PLASMA HALF-SPACE

Let an incident wave of frequency ω traveling in free space ($z < 0$) be incident normally on the plasma half-space ($z > 0$) of plasma frequency ω_p. The intrinsic impedance of the plasma medium is given by

$$\eta_p=\frac{\eta_0}{n_p}=\frac{120\pi\omega}{\sqrt{\omega^2-\omega_p^2}}. \tag{2.48}$$

From Equation 2.43 the reflection coefficient R_A is given by

$$R_A=\frac{\eta_p-\eta_0}{\eta_p+\eta_0}=\frac{1-n_p}{1+n_p}=\frac{\Omega-\sqrt{\Omega^2-1}}{\Omega+\sqrt{\Omega^2-1}}, \tag{2.49}$$

where

$$\Omega=\frac{\omega}{\omega_p} \tag{2.50}$$

is the source frequency normalized with respect to the plasma frequency. The power reflection coefficient $\rho = |R_A|^2$ and the power transmission coefficient τ ($= 1 - \rho$)

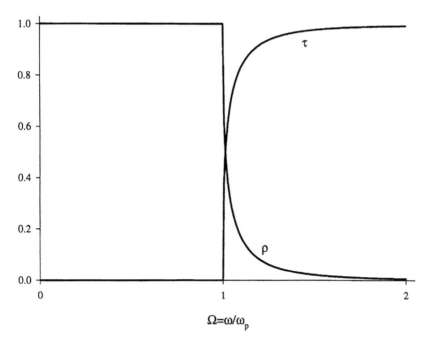

FIGURE 2.5 Sketch of the power reflection coefficient (ρ) and the power transmission coefficient (τ) of a plasma half-space.

vs. Ω are shown in Figure 2.5. In the frequency band $0 < \omega < \omega_p$, the characteristic impedance η_p is imaginary. The electric field $\mathbf{E}_p$ and the magnetic field $\mathbf{H}_p$ in the plasma are in *time quadrature*; the real part of $(\mathbf{E}_p \times \mathbf{H}_p^*)$ is zero. The wave in plasma is an *evanescent wave*[1] and carries no real power. The source wave is totally reflected and $\rho = 1$. In the frequency band $\omega_p < \omega < \infty$, the plasma behaves as a dielectric and the source wave is partially transmitted and partially reflected. The time-averaged power density of the incident wave is equal to the sum of the time-averaged power densities of the reflected and transmitted waves.

Metals at optical frequencies are modeled as plasmas with plasma frequency ω_p of the order of 10^{16}. Refining the plasma model by including the collision frequency will lead to three frequency domains of conducting, cutoff, and dielectric phenomena.[2] A more comprehensive account of modeling metals as plasmas is given in References 3 and 4. The associated phenomena of attenuated total reflection, surface plasmons, and other interesting topics are based on modeling metals as plasmas. A brief account of surface wave is given in Section 2.8.

2.4 REFLECTION BY A PLASMA SLAB

This section considers oblique incidence to add to the variety to the problem formulation. The geometry of the problem is shown in Figure 2.6. Let x–z be the plane of incidence and $\hat{k}$ be the unit vector along the direction of propagation. Let the magnetic field $\mathbf{H}^I$ be along y and the electric field $\mathbf{E}^I$ lie entirely in the plane of

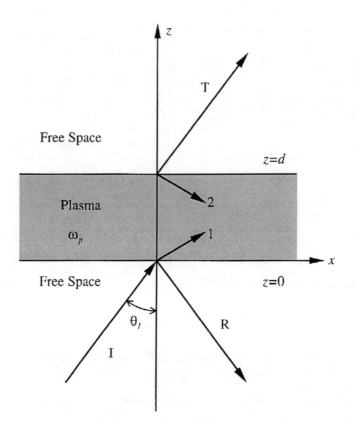

FIGURE 2.6   Plasma slab problem.

incidence. Such a wave is described in the literature by various names: (1) p wave, (2) TM wave, and (3) parallel polarized wave. Since the boundaries are at $z = 0$ and $z = d$, it is necessary to use x, y, z coordinates in formulating the problem and express $\hat{k}$ in terms of x and z coordinates. The problem therefore appears to be two dimensional. Equations 2.51 through 2.56 describe the incident wave:

$$\mathbf{H}^{\mathrm{I}} = \hat{y} H_y^{\mathrm{I}} \exp\left[-j k_0 \hat{k} \cdot \mathbf{r}\right], \tag{2.51}$$

$$\hat{k} = \hat{x} S + \hat{z} C, \tag{2.52}$$

$$S = \sin\theta_{\mathrm{I}}, \quad C = \cos\theta_{\mathrm{I}}, \tag{2.53}$$

$$\mathbf{E}^{\mathrm{I}} = \left(\hat{y} \times \hat{k}\right) E^{\mathrm{I}} \exp\left[-j k_0 \hat{k} \cdot \mathbf{r}\right] = \left(\hat{x} E_x^{\mathrm{I}} + \hat{z} E_z^{\mathrm{I}}\right) \Psi^{\mathrm{I}}, \tag{2.54}$$

$$E_x^{\mathrm{I}} = C E^{\mathrm{I}}, \quad E_z^{\mathrm{I}} = -S E^{\mathrm{I}}, \quad E^{\mathrm{I}} = \eta_0 H_y^{\mathrm{I}}, \tag{2.55}$$

$$\Psi^{I} = \exp\left[-jk_0(Sx + Cz)\right].$$ (2.56)

The reflected wave is written as

$$\mathbf{E}^{R} = \left(\hat{x}E_x^R + \hat{z}E_z^R\right)\Psi^R,$$ (2.57)

$$\mathbf{H}^{R} = \hat{y}H_y^R\Psi^R,$$ (2.58)

$$\Psi^{R} = \exp\left[-jk_0(Sx - Cz)\right].$$ (2.59)

The boundary conditions at $z = 0$ require that the x dependence in the region $z > 0$ remains the same as in the region $z < 0$. Since the region $z > d$ is also free space, the exponential factor for the transmitted wave will be the same as the incident wave and the fields of the transmitted wave are

$$\mathbf{E}^{T} = \left(\hat{x}E_x^T + \hat{z}E_z^T\right)\Psi^T,$$ (2.60)

$$\mathbf{H}^{T} = \hat{y}H_y^T\Psi^T,$$ (2.61)

$$\Psi^{T} = \exp\left[-jk_0(Sx + Cz)\right].$$ (2.62)

All the field amplitudes in free space can be expressed in terms of E_x^I, E_x^R, or E_x^T:

$$CE_z^{I,T} = -SE_x^{I,T},$$ (2.63)

$$CE_z^{R} = SE_x^{R},$$ (2.64)

$$C\eta_0 H_y^{I,T} = E_x^{I,T},$$ (2.65)

$$C\eta_0 H_y^{R} = -E_x^{R}.$$ (2.66)

Next the waves in the homogeneous plasma region $0 < z < d$ are considered. The x and the z dependence for waves in the plasma comes from the exponential factor Ψ^P:

$$\Psi^{P} = \exp\left[-jk_0(Sx + qz)\right],$$ (2.67)

where q has to be determined. The wave number in the plasma is given as $k_p = k_0\sqrt{\varepsilon_p}$. Thus,

$$q^2 + S^2 = \varepsilon_p = 1 - \frac{\omega_p^2}{\omega^2},$$

(2.68)

$$q_{1,2} = \pm\sqrt{C^2 - \frac{\omega_p^2}{\omega^2}} = \pm\sqrt{\frac{C^2\Omega^2 - 1}{\Omega^2}}.$$

(2.69)

The two values for q indicate the excitation of two waves in the plasma — the first wave is the positive wave and the second wave is a negative wave. For $\Omega < 1/C$, q is imaginary and the two waves in the plasma are evanescent. The next section deals with this frequency band.

The fields in the plasma can be written as

$$\mathbf{E}^P = \sum_{m=1}^{2} \left(\hat{x}E_{xm}^P + \hat{z}E_{zm}^P\right)\Psi_m^P,$$

(2.70)

$$\mathbf{H}^P = \sum_{m=1}^{2} \hat{y}H_{ym}^P \Psi_m^P,$$

(2.71)

$$\Psi_m^P = \exp\left[-jk_0\left(Sx + q_m z\right)\right].$$

(2.72)

All the field amplitudes in the plasma can be expressed in terms of E_{x1}^P and E_{x2}^P:

$$\eta_0 H_{ym}^P = \frac{\varepsilon_p}{q_m} E_{xm}^P = \eta_{ym} E_{xm}^P,$$

(2.73)

$$E_{zm}^P = -\frac{S}{q_m} E_{xm}^P.$$

(2.74)

The above relations are obtained from Equations 2.3 and 2.6 by noting $\partial/\partial x = -jk_0S$ and $\partial/\partial z = -jk_0q$; they may also be written from inspection. By assuming E_x^I is known, the unknowns reduce to four: E_{x1}^P, E_{x2}^P, E_x^R, and E_x^T. They can be determined from the four boundary conditions of continuity of the tangential components E_x and H_y at $z = 0$ and $z = d$. In matrix form

$$
\begin{bmatrix}
1 & 1 & -1 & 0 \\
C\eta_{y1} & C\eta_{y2} & 1 & 0 \\
\lambda_1 & \lambda_2 & 0 & -1 \\
\lambda_1 C\eta_{y1} & \lambda_2 C\eta_{y2} & 0 & -1
\end{bmatrix}
\begin{bmatrix}
E_{x1}^P \\
E_{x2}^P \\
E_x^R \\
E_x^T
\end{bmatrix}
=
\begin{bmatrix}
1 \\
1 \\
0 \\
0
\end{bmatrix}
E_x^I,
$$

(2.75)

where

$$\lambda_1 = \exp\left[jk_0(C - q_1)d\right] \quad \text{and}$$
$$\lambda_2 = \exp\left[jk_0(C - q_2)d\right]. \tag{2.76}$$

By solving Equation 2.75,

$$E_{x1}^P = 2\lambda_2\left(1 - C\eta_{y2}\right)E_x^I/\Delta, \tag{2.77}$$

$$E_{x2}^P = -2\lambda_1\left(1 - C\eta_{y1}\right)E_x^I/\Delta, \tag{2.78}$$

$$E_x^R = \left(1 - C\eta_{y1}\right)\left(1 - C\eta_{y2}\right)\left(\lambda_2 - \lambda_1\right)E_x^I/\Delta, \tag{2.79}$$

$$E_x^T = 2\lambda_1\lambda_2 C\left(\eta_{y1} - \eta_{y2}\right)E_x^I/\Delta, \tag{2.80}$$

where

$$\Delta = \lambda_2\left(1 + C\eta_{y1}\right)\left(1 - C\eta_{y2}\right) - \lambda_1\left(1 - C\eta_{y1}\right)\left(1 + C\eta_{y2}\right). \tag{2.81}$$

The power reflection coefficient $\rho = \left|E_x^R / E_x^I\right|^2$ is given by

$$\rho = \frac{1}{1 + \left(\dfrac{2C\varepsilon_p q_1}{C^2\varepsilon_p^2 - q_1^2} \operatorname{cosec}\left(2\pi\Omega q_1 d_p\right)\right)^2}, \tag{2.82}$$

where

$$d_p = \frac{d}{\lambda_p}, \tag{2.83}$$

$$\lambda_p = \frac{2\pi c}{\omega_p}. \tag{2.84}$$

In the above, λ_p is the free-space wavelength corresponding to the plasma frequency and is used to normalize the slab width. By substituting for q_1, in the range of real q_1

$$\rho = \frac{1}{1 + B\cosec^2 A}, \quad \frac{1}{C} < \Omega < \infty \tag{2.85}$$

and in the range of imaginary q_1

$$\rho = \frac{1}{1 - B\cosech^2|A|}, \quad 0 < \Omega < \frac{1}{C}, \tag{2.86}$$

where A and B are given by

$$A = 2\pi\Omega q_1 d_p = 2\pi\sqrt{C^2\Omega^2 - 1}\, d_p \quad \text{and} \tag{2.87}$$

$$B = \frac{4C^2\varepsilon_p^2 q_1^2}{\left(C^2\varepsilon_p^2 - q_1^2\right)^2} = \frac{4C^2\Omega^2\left(\Omega^2 - 1\right)^2\left(C^2\Omega^2 - 1\right)}{\left(2C^2\Omega^2 - \Omega^2 - C^2\right)^2}. \tag{2.88}$$

The power transmission coefficient $\tau = (1 - \rho)$ and is given by

$$\tau = \frac{|B|}{|B| + \sin^2|A|}, \quad \frac{1}{C} < \Omega < \infty \quad \text{and} \tag{2.89}$$

$$\tau = \frac{|B|}{|B| + \sinh^2|A|}, \quad 0 < \Omega < \frac{1}{C}. \tag{2.90}$$

From Equation 2.89, $\rho = 0$ when $A = 0$ or

$$\sin A = n\pi, \quad n = 0, 1, 2, \ldots \tag{2.91}$$

There is one more value of Ω for which $\rho = 0$. From Equation 2.88, when $C\varepsilon_p = q_1$, $B = \infty$, and from Equation 2.89 $\tau = 1$ and $\rho = 0$. This point corresponds to a frequency Ω_B given by

$$\Omega_B^2 = \frac{C^2}{2C^2 - 1}. \tag{2.92}$$

It is easily shown that Ω_B exists only for the p wave and is greater than $1/C$. In fact, this point corresponds to the *Brewster angle*.[4] Figure 2.7 shows Ω_B vs. $\cos\theta_B$, where θ_B is the Brewster angle.

Thus, in the frequency band $\Omega > 1/C$, the variation of ρ with the slab width is oscillatory. Stratton[6] discussed these oscillations for a dielectric slab and associated them with the interference of the internally reflected waves in the slab. In the case

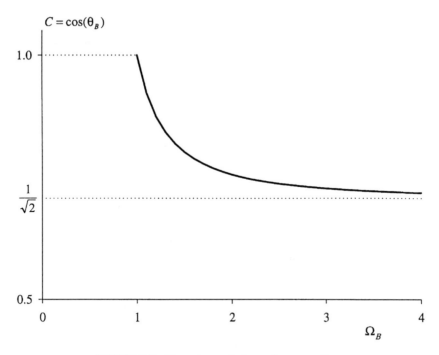

FIGURE 2.7 Brewster angle for a plasma medium.

of the plasma slab, the variation of ρ with the source frequency is also oscillatory, since in this range the plasma behaves like a dispersive dielectric. The maxima of ρ are less than or equal to 1 and occur for Ω satisfying the transcendental equation

$$\tan A = fA, \tag{2.93}$$

where

$$f = \frac{C^2\Omega^2(\Omega^2 - 1)(C^2 + \Omega^2 - 2C^2\Omega^2)}{(2C^2\Omega^4 - 2C^2\Omega^2 + C^2 - \Omega^2)(1 + \Omega^2 - 2C^2\Omega^2)}. \tag{2.94}$$

Figure 2.8 shows ρ vs. Ω for a normalized slab width $d_p = 1.0$ and $C = 0.5(\theta_1 = 60°)$. The inset shows the variation of q with Ω. The oscillations can be clearly seen in the real range of q. Reference 2 shows similar curves for normal incidence on a lossy plasma slab using parameters normalized with reference to the source wave quantities.

In Figure 2.8, $\rho \approx 1$ in the imaginary range of q showing total reflection of the source wave. However, it will be shown that there can be considerable *tunneling of power* for sufficiently thin plasma slabs.

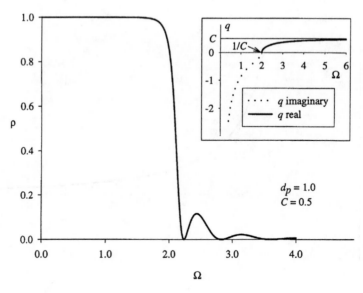

FIGURE 2.8 ρ vs. Ω for isotropic plasma slab (parallel polarization).

2.5   TUNNELING OF POWER THROUGH A PLASMA SLAB

For the frequency band $\Omega < 1/C$, the characteristic roots are imaginary and the waves excited in the plasma are evanescent. The incident wave is completely reflected by the semi-infinite plasma. However, in the case of a plasma slab, some power gets transmitted through it even in this frequency band. It can be shown that this tunneling effect is due to the interaction of the electric field of the positive wave with the magnetic field of the negative wave and vice versa.

The above statement is supported by examining the power flow through the Poynting vector calculation. From Equations 2.77 through 2.81, one can obtain

$$E_x^{\mathrm{T}} = 2C\varepsilon_p q_1 \exp\left[jk_0 Cd\right] E_x^{\mathrm{I}}/\Delta, \tag{2.95}$$

$$E_{x1}^{\mathrm{P}} = \left(q_1 + C\varepsilon_p\right)q_1 \exp\left[jk_0 q_1 d\right] E_x^{\mathrm{I}}/\Delta, \tag{2.96}$$

$$E_{x2}^{\mathrm{P}} = -\left(-q_1 + C\varepsilon_p\right)q_1 \exp\left[-jk_0 q_1 d\right] E_x^{\mathrm{I}}/\Delta, \tag{2.97}$$

where

$$\Delta = 2q_1 C\varepsilon_p \cos\left(k_0 q_1 d\right) + j\left(q_1^2 + C^2\varepsilon_p^2\right)\sin\left(k_0 q_1 d\right). \tag{2.98}$$

The magnetic field components H_{y1}^{P}, H_{y2}^{P}, and H_y^{T} can be obtained from Equations 2.73 and 2.65.

The power crossing any plane in the plasma parallel to the interface can be found by calculating the z-component of the Poynting vector associated with the plasma waves. This is given by

$$S_x^P = \tfrac{1}{2}\mathrm{Re}\Big[\big(E_{x1}^P\Psi_1^P + E_{x2}^P\Psi_2^P\big)\big(H_{y1}^{P*}\Psi_1^{P*} + H_{y2}^{P*}\Psi_2^{P*}\big)\Big], \qquad (2.99)$$

which on expansion gives

$$S_z^P = \tfrac{1}{2}\mathrm{Re}\Big[E_{x1}^P\Psi_1^P H_{y1}^{P*}\Psi_1^{P*}\Big] + \tfrac{1}{2}\mathrm{Re}\Big[E_{x2}^P\Psi_2^P H_{y1}^{P*}\Psi_1^{P*}\Big]$$
$$+\ \tfrac{1}{2}\mathrm{Re}\Big[E_{x1}^P\Psi_1^P H_{y2}^{P*}\Psi_2^{P*}\Big] + \tfrac{1}{2}\mathrm{Re}\Big[E_{x2}^P\Psi_2^P H_{y2}^{P*}\Psi_2^{P*}\Big], \qquad (2.100)$$

where Ψ_1^P and Ψ_2^P are given by Equation 2.72. The contribution from each of the four terms on the right side of Equation 2.100 is now discussed for the two cases of q_1 real $\Omega > 1/C$ and q_1 imaginary $\Omega > 1/C$. It is to be noted that the third and the fourth terms give the contributions that are due to the cross interaction of the fields in the positive-going and negative-going waves.

Case 1: q_1 Real. It is easy to see that because the third and the fourth terms are equal but opposite in sign, there is no contribution to the net power flow from the cross interaction. The sum of the first and second terms is equal to $\tfrac{1}{2}\mathrm{Re}[(E_x^T\Psi^T)(H_y^{T*})]$, where Ψ^T is given by Equation 2.62. Thus, the power crossing any plane parallel to the slab interface gives exactly the power that emerges into free space at the other boundary of the slab.

Case 2: q_1 Imaginary. It can be shown that each of the first two terms is zero. This is because the corresponding electric and magnetic fields are in time quadrature. Furthermore, the third term is equal to the fourth term, and each is equal to $\tfrac{1}{4}\mathrm{Re}[(E_x^T\Psi^T)(H_y^{T*}\Psi^{T*})]$. Thus, the power flow in the tunneling frequency band ($\Omega<1/C$) comes entirely from the cross interaction.

At $\Omega = 1$, $\varepsilon_p = 0$ and from Equation 2.88, $B = 0$ but $A \neq 0$. From Equation 2.89 $\tau = 0$. This point exists only for the p wave.

At $\Omega = 1/C$, $A = 0$ and $B = 0$. By evaluating the limit, the power transmission coefficient τ_c at this point is obtained:

$$\tau_c = \frac{1}{1+S^4\pi^2 d_p^2}, \quad \Omega = \frac{1}{C}. \qquad (2.101)$$

Numerical results of τ vs. Ω are presented in Figures 2.9 and 2.10 for the tunneling frequency band. In Figure 2.9, the angle of incidence is taken to be 60°

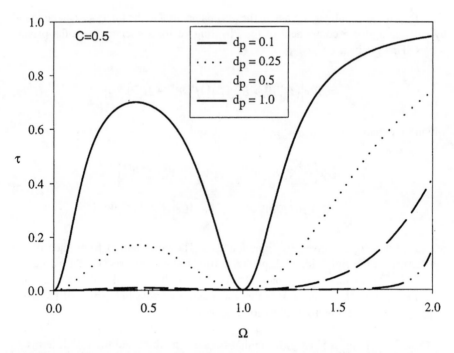

FIGURE 2.9 Transmitted power for an isotropic plasma slab (parallel polarization) in the tunneling range ($\Omega < 1/C$) for various d_p.

and the curves are given for various d_p. Between $\Omega = 0$ and 1, there is a value of $\Omega = \Omega_{max}$, where the transmission is maximum. The maximum value of the transmitted power decreases as d_p increases.

In Figure 2.10, τ vs. Ω is shown with the angle of incidence as the parameter. It is seen that the point of maximum transmission moves to the right as the angle of incidence decreases and merges with $\Omega = 1$ for $\theta_I = 0°$ (normal incidence). The maximum value of the transmitted power decreases as θ_I increases. This point is given by the solution of the transcendental equation

$$\tanh|A| = f|A|, \qquad (2.102)$$

where f and A are given in Equations 2.94 and 2.87, respectively. It is suggested that by measuring Ω_{max} experimentally either the plasma frequency or the slab width can be determined if the angle of incidence is known.

2.6   INHOMOGENEOUS SLAB PROBLEM

The case where ε_p is a function of z in the region $0 < z < d$ is considered next, by reverting back to the normal incidence to simplify and focus on the effect of the inhomogeneity of the properties of the medium. The differential equation for the

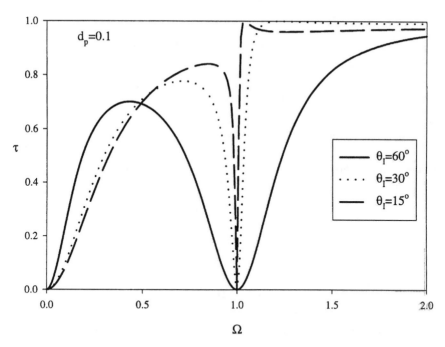

FIGURE 2.10 Transmitted power for an isotropic plasma slab (parallel polarization) in the tunneling range ($\Omega < 1/C$) for various angles of incidence.

electric field is given by Equation 2.15. This equation cannot be solved exactly for a general $\varepsilon_p(z)$ profile.

However, there are a small number of profiles for which exact solutions can be obtained in terms of *special functions*. As an example, the problem of the *linear profile* can be solved in terms of *Airy functions*. An account of such solutions for a dielectric profile $\varepsilon_r(z)$ is given in References 8 and 9 and for a plasma profile $\varepsilon_p(z, \omega)$ in References 2 and 10. Reference 11 gives detailed information on special functions.

A vast amount of literature is available on the approximate solution for an inhomogeneous media problem. The following techniques are emphasized in this book. The actual profile is considered as a perturbation of a profile for which the exact solution is known. The effect of the perturbation is calculated through the use of a Green's function. The problem of a *fast profile* with finite range length can be so handled by using a step profile as the reference. At the other end of the approximation scale, the *slow profile* problem can be handled by adiabatic and Wentzel, Krammers, and Brillouin (WKB) techniques. See References 1 and 8 for dielectric profile examples and References 2 and 10 for plasma profile examples. Numerical approximation techniques can also be used, which are extremely useful in obtaining specific numbers for a given problem and in validating the theoretical results obtained from the physical approximations. The subsequent chapters discuss these aspects for temporal profiles.

2.7   PERIODIC LAYERS OF PLASMA

A particular case of an inhomogeneous plasma medium that is periodic is considered next. Let the dielectric function

$$\varepsilon_p(z,\omega) = \varepsilon_p(z+mL,\omega), \tag{2.103}$$

where m is an integer and L is the spatial period. If the domain of m is all integers, $-\infty < m < \infty$, an unbounded periodic media exists. The solution of Equation 2.15 for the infinite periodic structure problem[1,9,12] will be such that the electric field $E(z)$ differs from the electric field at $E(z + L)$ by a constant:

$$E(z+L) = CE(z). \tag{2.104}$$

The complex constant C can be written as

$$C = \exp(-j\beta L), \tag{2.105}$$

where β is the complex propagation constant of the periodic media. Thus,

$$E(z) = E_\beta(z)\exp(-j\beta z). \tag{2.106}$$

From Equation 2.106,

$$E(z+L) = E_\beta(z+L)\exp(-j\beta(z+L)). \tag{2.107}$$

Also, from Equations 2.104 and 2.106,

$$E(z+L) = E(z)\exp(-j\beta L) = E_\beta(z)\exp(-j\beta z)\exp(-j\beta L)$$
$$= E_\beta(z)\exp(-j\beta(z+L)). \tag{2.108}$$

From Equations 2.107 and 2.108, it follows that $E_\beta(z)$ is periodic:

$$E_\beta(z+L) = E_\beta(z). \tag{2.109}$$

Equation 2.106, where $E_\beta(z)$ is periodic, is called the Bloch wave condition. Such a periodic function can be expanded in a Fourier series:

$$E_\beta(z) = \sum_{m=-\infty}^{\infty} A_m \exp(-j2m\pi/L), \tag{2.110}$$

and $E(z)$ can be written as

$$E(z) = \sum_{m=-\infty}^{\infty} A_m \exp\left[-j(\beta + 2m\pi/L)z\right] = \sum_{m=-\infty}^{\infty} A_m \exp\left(-j\beta_m z\right), \qquad (2.111)$$

where

$$\beta_m = \beta + \frac{2m\pi}{L}. \qquad (2.112)$$

By taking into account both positive-going and negative-going waves, $E(z)$ can be expressed as[1]

$$E(z) = \sum_{m=-\infty}^{\infty} A_m \exp\left(-j\beta_m z\right) + \sum_{m=-\infty}^{\infty} B_m \exp\left(+j\beta_m z\right). \qquad (2.113)$$

The wave propagation in a periodic layered media with plasma layers alternating with free space is considered next. The geometry of the problem is shown in Figure 2.11. By adopting the notation of Reference 12, a unit cell consists of free space from $-l < z < l$ and a plasma layer of plasma frequency ω_{p0} from $l < z < l + d$. The thickness of the unit cell is $L = 2l + d$. The electric field $E(z)$ in the two layers of the unit cell can be written as

$$E(z) = A \exp\left(-j\frac{\omega}{c}z\right) + B \exp\left(j\frac{\omega}{c}z\right), -l \le z \le l, \qquad (2.114)$$

$$C \exp\left[-jn\frac{\omega}{c}(z-l)\right] + D \exp\left[jn\frac{\omega}{c}(z-l)\right], l \le z \le l + d, \qquad (2.115)$$

where n is the refractive index of the plasma medium:

$$n = \sqrt{\varepsilon_p} = \sqrt{1 - \omega_{p0}^2/\omega^2}. \qquad (2.116)$$

The continuity of tangential electric and magnetic fields at the boundary translates into the continuity of E and $\partial E/\partial z$ at the interfaces:

$$E(l^-) = E(l^+), \qquad (2.117)$$

$$\frac{\partial E}{\partial z}(l^-) = \frac{\partial E}{\partial z}(l^+), \qquad (2.118)$$

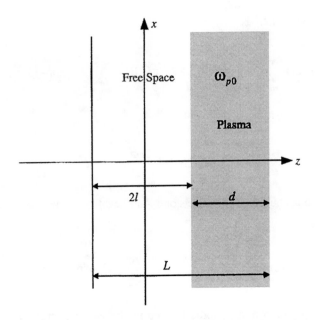

FIGURE 2.11 Unit cell of unbounded periodic media consisting of layer of plasma.

$$E(l+d^-) = E(l+d^+),$$ (2.119)

$$\frac{\partial E}{\partial z}(l+d^-) = \frac{\partial E}{\partial z}(l+d^+).$$ (2.120)

From Equations 2.117 and 2.118,

$$A\exp\left(-j\frac{\omega}{c}L\right) + B\exp\left(j\frac{\omega}{c}L\right) = C+D,$$ (2.121)

$$A\exp\left(-j\frac{\omega}{c}L\right) - B\exp\left(j\frac{\omega}{c}L\right) = n(C-D).$$ (2.122)

Since $l + d^+ = -l + 2d^+ = -l^+ + L$, from Equation 2.104

$$E(l+d^+) = E(-l^+ + L) = \exp(-j\omega L)E(-l^+).$$ (2.123)

From Equations 2.118, 2.119, 2.113, and 2.114,

$$C \exp\left(-jn\frac{\omega}{c}d\right) + D \exp\left(jn\frac{\omega}{c}d\right) =$$
$$\exp(-j\beta L)\left[A \exp\left(-j\frac{\omega}{c}l\right) + B \exp\left(j\frac{\omega}{c}l\right)\right].$$

(2.124)

From Equations 2.120, 2.115, 2.123, and 2.114,

$$-n\frac{\omega}{c}C \exp\left(-jn\frac{\omega}{c}d\right) + n\frac{\omega}{c}D \exp\left(jn\frac{\omega}{c}d\right)$$
$$= \exp(-j\beta L)\left[-j\frac{\omega}{c}A \exp\left(-\frac{j\omega l}{c}\right) + j\frac{\omega}{c}B \exp\left(j\frac{\omega}{c}l\right)\right].$$

(2.125)

Equations 2.121, 2.122, 2.124, and 2.125 may be arranged as a matrix:

$$\begin{bmatrix} e^{-j\omega l/c} & e^{+j\omega l/c} & -1 & -1 \\ e^{-j\omega l/c} & -e^{j\omega l/c} & -n & n \\ e^{+j\omega l/c} & e^{-j\omega l/c} & -e^{j(\beta L-n\omega d/c)} & -e^{j(\beta L+n\omega d/c)} \\ e^{+j\omega l/c} & -e^{-j\omega l/c} & -ne^{j(\beta L-n\omega d/c)} & ne^{j(\beta L+n\omega d/c)} \end{bmatrix} \begin{bmatrix} A \\ B \\ C \\ D \end{bmatrix} = 0 \quad (2.126)$$

A nonzero solution for the fields can be obtained by equating the determinant of the square matrix in Equation 2.126 to zero. This leads to the dispersion relation:

$$\cos\beta L = \cos\frac{n\omega d}{c}\cos\frac{2\omega l}{c} - \frac{1}{2}\left[n+\frac{1}{n}\right]\sin\frac{n\omega d}{c}\sin\frac{2\omega l}{c}.$$

(2.127)

The above equation can be studied by using β_r and $\omega_r = \beta_r c = c/2\pi\lambda_r$ as reference values. Thus, Equation 2.127, in the normalized form, can be written as

$$\cos\left(\frac{\beta}{\beta_r}\frac{2\pi L}{\lambda_r}\right) = \cos\left(\frac{n\omega}{\omega_r}\frac{2\pi d}{\lambda_r}\right)\cos\left(\frac{2\omega}{\omega_r}\frac{2\pi L}{\lambda_r}\right)$$
$$-\frac{1}{2}\left(n+\frac{1}{n}\right)\sin\left(\frac{n\omega}{\omega_r}\frac{2\pi d}{\lambda_r}\right)\sin\left(\frac{2\omega}{\omega_r}\frac{2\pi L}{\lambda_r}\right).$$

(2.128)

Figure 2.12 shows the graph ω/ω_r vs. β/β_r for the following values of the parameters: $L = 0.6\lambda_r$, $d = 0.2\lambda_r$, and $\omega_{p0} = 1.2\omega_r$. It is noted from Figure 2.12 that the wave is evanescent in the frequency band $0 < \omega/\omega_r < 0.611$. This stop band is due to the plasma medium in the layers. If the layers are dielectric, this stop band will not be

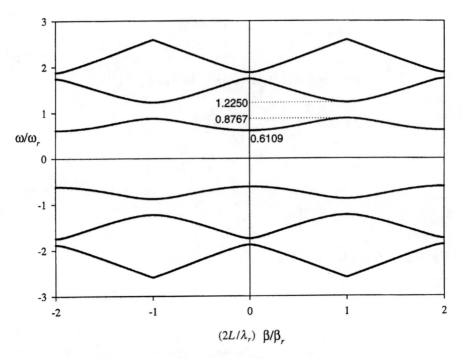

FIGURE 2.12 Dispersion relation of periodic plasma medium for $d = 0.2\lambda$, $L = 0.6\lambda$.

present. There is again a stop band in the frequency domain $0.877 < \omega/\omega_r < 1.225$, and this stop band is due to the periodicity of the medium. Periodic dielectric layers do exhibit such a forbidden band. This principle is used in optics to construct *Bragg reflectors* with extremely large reflectance.[9] Figure 2.13 shows a $\omega - \beta$ diagram for dielectric layers with the refractive index $n = 1.8$. The first stop band in this case is given by $0.537 < \omega/\omega_r < 0.779$.

2.8   SURFACE WAVES

It is necessary to backtrack a little bit and consider the oblique incidence of a p wave on a plasma half-space. In this case only the outgoing wave will be excited in the plasma half-space, and Equation 2.75 becomes

$$\begin{bmatrix} 1 & -1 \\ c\eta_{y1} & 1 \end{bmatrix} \begin{bmatrix} E_{x1}^P \\ E_x^R \end{bmatrix} = \begin{bmatrix} 1 \\ 0 \end{bmatrix} E_x^I. \tag{2.129}$$

Solution of Equation 2.129 gives the fields of the reflected wave and the wave in the plasma half-space in terms of the fields of the incident wave. If E_x^I is zero (no incident wave), $E_{x1}^P = E_x^R = 0$ is expected. This is true in general with one exception. The exception occurs when the determinant of the square matrix on the left side of Equation 2.129 is zero. In such a case E_{x1}^P and E_x^R may be nonzero even if E_x^I is zero,

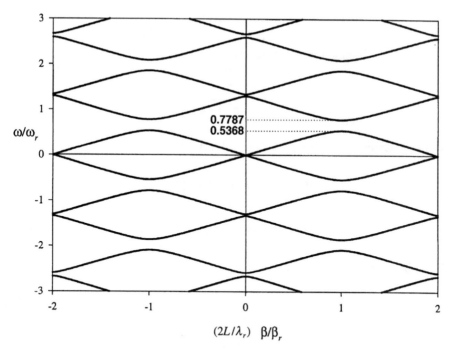

FIGURE 2.13   Dispersion relation of periodic dielectric layers for $d = 0.2\lambda$, $L = 0.6\lambda$. The refractive index of the dielectric layer is 1.8.

indicating the possibility of the existence of fields in free space even if there is no incident field. For the exceptional solution, the reflection coefficient is infinity, i.e., $E_x^R \neq 0$, but $E_x^I = 0$. The solution is an eigenvalue solution and gives the dispersion relation for surface plasmons. From Equations 2.129 and 2.73 the dispersion relation is obtained as

$$1 + C\eta_{y1} = 1 + C\frac{\varepsilon_p}{q_1} = 0. \tag{2.130}$$

By noting that

$$k_0 q_1 = \sqrt{k_0^2 \varepsilon_p - k_x^2} \tag{2.131}$$

and

$$k_0^2 C^2 = \left(k_0^2 - k_x^2\right) \tag{2.132}$$

where

$$k_0 = \frac{\omega}{c}, \tag{2.133}$$

$$k_x = k_0 S, \tag{2.134}$$

the dispersion relation can be written as

$$k_x^2 = \frac{\omega^2}{c^2} \frac{\varepsilon_p}{1 + \varepsilon_p}. \tag{2.135}$$

The exponential factors for $z < 0$ and $z > 0$ can be written as

$$\Psi^{P1} = \exp\left[-jk_x x - jk_0 q_1 z\right] \quad z > 0, \tag{2.136}$$

$$\Psi^R = \exp\left[-jk_x x + jk_0 C z\right] \quad z < 0. \tag{2.137}$$

It can be shown that k_x will be real and both

$$\alpha_1 = jk_0 q_1 \tag{2.138}$$

and

$$\alpha_2 = jk_0 C = \sqrt{k_0^2 - k_x^2} \tag{2.139}$$

will be real and positive if

$$\varepsilon_p < -1. \tag{2.140}$$

When Equation 2.140 is satisfied,

$$\Psi^{P1} = \exp\left(-\alpha_1 z\right)\exp\left(-jk_x x\right), \quad z > 0, \tag{2.141}$$

$$\Psi^R = \exp\left(\alpha_2 z\right)\exp\left(jk_x x\right), \quad z < 0, \tag{2.142}$$

where α_1 and α_2 are positive real quantities. Equations 2.141 and 2.142 show that the waves, while propagating along the surface, attenuate in the direction normal to the surface. For this reason, the wave is called a surface wave. In a plasma when $\omega < \omega_p / \sqrt{2}$, ε_p will be less than -1. Thus, an interface between free space and a plasma medium can support a surface wave.

By defining the refractive index of the surface mode as n, i.e.,

$$n = \frac{c}{\omega} k_x, \tag{2.143}$$

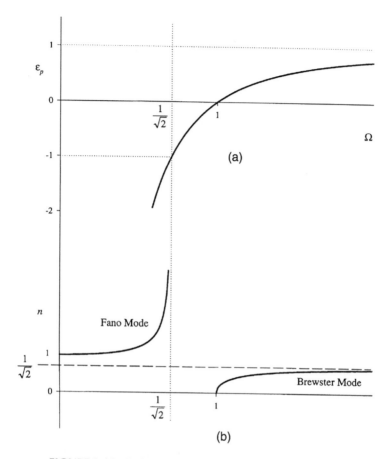

FIGURE 2.14 Refractive index for surface plasmon modes.

Equation 2.135 can be written as

$$n^2 = \frac{\Omega^2 - 1}{2(\Omega^2 - 1/2)}, \Omega < \frac{1}{\sqrt{2}}.$$ (2.144)

For $0 < \omega < \omega_p/\sqrt{2}$, n is real and the surface wave propagates. In this interval of ω, the dielectric constant $\varepsilon_p < -1$. The surface wave mode is referred to in the literature as the Fano mode[13] and also as a nonradiative surface plasmon. In the interval $1 < \Omega < \infty$, n^2 is positive and less than 0.5. However, the α values obtained from Equations 2.138 and 2.139 are imaginary. The wave is not bound to the interface, and the mode, called the Brewster mode, in this case is radiative and referred to as a radiative surface plasmon. Figure 2.14 sketches $\varepsilon_p(\omega)$ and $n(\omega)$. Another way of presenting the information is through a Ω–K diagram, where Ω and K are normalized frequency and wave number, respectively, i.e.,

$$\Omega = \frac{\omega}{\omega_p},$$ (2.145)

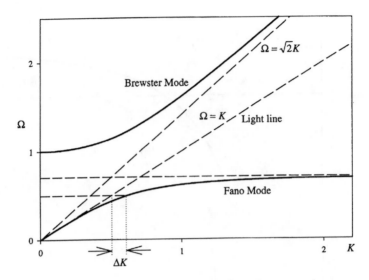

FIGURE 2.15 $\Omega - K$ diagram for surface plasmon modes.

$$K = \frac{ck_x}{\omega_p} = n\Omega. \tag{2.146}$$

From Equations 2.143 and 2.144, the following relation is obtained:

$$K^2 = n^2\Omega^2 = \frac{\Omega^2\left(\Omega^2 - 1\right)}{2\Omega^2 - 1}. \tag{2.147}$$

Figure 2.15 shows the Ω–K diagram.

The Fano mode is a surface wave called the surface plasmon and exists for $\omega <$ $\omega_p /\sqrt{2}$. The surface mode does not have a low-frequency cutoff, it is guided by an open-boundary structure, and has a phase velocity less than that of light. The eigenvalue spectrum is continuous, which is in contrast to the eigenvalue spectrum of a conducting-boundary waveguide that has an infinite number of discrete modes of propagation at a given frequency.

The application of photons (electromagnetic light waves) to excite the surface plasmons meets with the difficulty that the dispersion relation of the Fano mode lies to the right of the light line $\Omega = K$ (see Figure 2.15). At a given photon energy $\hbar\omega$ the wave vector $\hbar(\omega/c)$ has to be increased by $\Delta k_x = (\omega_p/c)\,\Delta k$ to transform the photons into surface plasmons. The two techniques commonly used to excite surface plasmons are (1) the grating coupler and (2) the attenuated total reflection (ATR) method. A complete account of surface plasmons can be found in References 13 and 14.

Surface waves can also exist at the interface of a dielectric with a plasma layer or a plasma cylinder. The theory of surface waves on a gas-discharge plasma is given in Reference 15 and Kalluri[16] used this theory to explore the possibility of backscatter from a plasma plume. Moissan et al.[17] achieved plasma generation through surface plasmons in a device called *Surfatron*; which is a highly efficient device for launching surface plasmons that produce plasma at a microwave frequency.

REFERENCES

1. Ishimaru, A., *Electromagnetic Wave Propagation, Radiation, and Scattering*, Prentice-Hall, Englewood Cliffs, NJ, 1991.
2. Heald, M. A. and Wharton, C. B., *Plasma Diagnostics with Microwaves*, Wiley, New York, 1965.
3. Forstmann, F. and Gerhardts, R. R., *Metal Optics Near the Plasma Frequency*, Springer-Verlag, New York, 1986.
4. Boardman, A. D., *Electromagnetic Surface Modes*, John Wiley, New York, 1982.
5. Kalluri, D. K. and Prasad, R. C., Thin film reflection properties of isotropic and uniaxial plasma slabs, *Appl. Sci. Res.* (Netherlands), 27, 415, 1973.
6. Stratton, J. A., *Electromagnetic Theory*, McGraw-Hill, New York, 1941.
7. Kalluri, D. K. and Prasad, R. C., Transmission of power by evanescent waves through a plasma slab, 1980 *IEEE International Conference on Plasma Science*, Madison, 11, 1980.
8. Lekner, J., *Theory of Reflection*, Kluwer, Boston, 1987.
9. Yeh, P., *Optical Waves in Layered Media*, John Wiley, New York, 1988.
10. Budden, K. G., *Radio Waves in the Ionosphere*, Cambridge University Press, Cambridge, 1961.
11. Abramowitz, M. and Stegun, I. A., *Handbook of Mathematical Functions*, Dover Publications, New York, 1965.
12. Kuo, S. P. and Faith, J., Interaction of an electromagnetic wave with a rapidly created spatially periodic plasma, *Phys. Rev. E*, 56, 1, 1997.
13. Boardman, A. D., Hydrodynamic theory of plasmon-polaritons on plane surfaces, in *Electromagnetic Surface Modes*, Boardman, A. D., Ed., John Wiley, New York, 1982, chap. 1.
14. Raether, H., *Surface Plasmons*, Springer-Verlag, New York, 1988.
15. Shivarova, A. and Zhelyazkov, I., Surface waves in gas-discharge plasmas, in *Electromagnetic Surface Modes*, Boardman, A. D., Ed., John Wiley, New York, 1982, chap. 12.
16. Kalluri, D. K., Backscattering from a Plasma Plume due to Excitation of Surface Waves, Final Report Summer Faculty Research Program, Air Force Office of Scientific Research, 1994.
17. Moissan, C., Beandry, C., and Leprince, P., *Phys. Lett.*, 50A, 125, 1974.

3 Time-Varying and Space-Invariant Isotropic Plasma Medium

3.1 BASIC EQUATIONS

If it is assumed that electron density varies only in time:

$$\omega_p^2 = \omega_p^2(t), \tag{3.1}$$

then the basic solutions can be constructed by assuming that the field variables vary harmonically in space:

$$\mathbf{F}(\mathbf{r}, t) = \mathbf{F}(t) \exp(-j\mathbf{k} \cdot \mathbf{r}). \tag{3.2}$$

Equations 1.1, 1.2, and 1.10 then reduce to Equations 3.3 through 3.5:

$$\mu_0 \frac{\partial \mathbf{H}}{\partial t} = j\mathbf{k} \times \mathbf{E}, \tag{3.3}$$

$$\varepsilon_0 \frac{\partial \mathbf{E}}{\partial t} = -j\mathbf{k} \times \mathbf{H} - \mathbf{J}, \tag{3.4}$$

$$\frac{d\mathbf{J}}{dt} = \varepsilon_0 \omega_p^2(t) \mathbf{E}. \tag{3.5}$$

From Equations 3.4 and 3.5 or from Equation 1.11, the wave equation for $\mathbf{E}$ is obtained:

$$\frac{d^2\mathbf{E}}{dt^2} + \left[k^2 c^2 + \omega_p^2(t)\right]\mathbf{E} - \mathbf{k}(\mathbf{k} \cdot \mathbf{E}) = 0. \tag{3.6}$$

From Equation 1.15

$$\frac{d^2\dot{\mathbf{H}}}{dt^2} + \left[k^2 c^2 + \omega_p^2(t)\right]\dot{\mathbf{H}} = 0. \tag{3.7}$$

The one-dimensional Equations 1.20 through 1.25 reduce to

$$\mu_0 \frac{\partial H}{\partial t} = jkE, \tag{3.8}$$

$$\varepsilon_0 \frac{\partial E}{\partial t} = jkH - J, \tag{3.9}$$

$$\frac{dJ}{dt} = \varepsilon_0 \omega_p^2(t)E, \tag{3.10}$$

$$\frac{d^2E}{dt^2} + \left[k^2c^2 + \omega_p^2(t)\right]E = 0, \tag{3.11}$$

$$\frac{d^2\dot{H}}{dt^2} + \left[k^2c^2 + \omega_p^2(t)\right]\dot{H} = 0. \tag{3.12}$$

If H rather than $\dot{H}$ is considered as the dependent variable, Equation 3.12 becomes

$$\frac{d^3H}{dt^3} + \left[k^2c^2 + \omega_p^2(t)\right]\frac{dH}{dt} = 0. \tag{3.13}$$

The solution of Equations 3.11 and 3.13, when ω_p^2 is a constant, can be easily obtained as

$$E(t) = E_0 \sum_{m=1}^{3} E_m \exp\left(j\omega_m t\right), \tag{3.14}$$

$$H(t) = H_0 \sum_{m=1}^{3} H_m \exp\left(j\omega_m t\right), \tag{3.15}$$

where

$$\omega_{1,2} = \pm\sqrt{k^2c^2 + \omega_p^2} \tag{3.16}$$

and

$$\omega_3 = 0. \tag{3.17}$$

In the above, E_0 and H_0 are the initial values of the electric and magnetic fields. E_m and H_m can be determined by solving the initial value problem. A detailed discussion

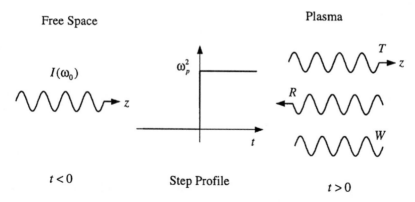

FIGURE 3.1 Suddenly created unbounded plasma medium.

of such a solution is given in the next section by considering a canonical problem of the sudden switching of an unbounded homogeneous plasma medium.

3.2 REFLECTION BY A SUDDENLY CREATED UNBOUNDED PLASMA MEDIUM

The geometry of the problem is shown in Figure 3.1. A plane wave of frequency Ω_0 is propagating in free space in the z-direction. Suddenly, at $t = 0$ a plasma medium is created. Thus, a temporal discontinuity in the dielectric properties of the medium is created. Let the exponential wave function of the source wave, reflected wave, and the transmitted wave be given by

$$\Psi^I = \exp\left[j\left(\omega_0 t - k_0 z\right)\right],\tag{3.18}$$

$$\Psi^R = \exp\left[j\left(-\omega_r t - k_r z\right)\right],\tag{3.19}$$

$$\Psi^T = \exp\left[j\left(\omega_t t - k_t z\right)\right].\tag{3.20}$$

In implementing the initial conditions of the continuity of the electric and magnetic fields at $t = 0$ for all z, the same coefficient of z in Equations 3.18 through 3.20 is needed, giving the condition:

$$k_0 = k_r = k_t = k.\tag{3.21}$$

In the case of temporal discontinuity, the wave number is conserved. The free-space wave number is given by

$$k_0 = \omega_0/c\tag{3.22}$$

and the wave number in the plasma is given by

$$k_p = \frac{\omega}{c}\sqrt{\varepsilon_p} = \sqrt{\omega^2 - \omega_p^2},$$ (3.23)

where ω is the frequency of the waves in the plasma. From Equations 3.21 through 3.23, $\omega_0^2 = \omega^2 - \omega_p^2$, leading to

$$\omega = \pm\sqrt{\omega_0^2 + \omega_p^2}.$$ (3.24)

The positive sign gives the frequency of the transmitted wave* ω_t and the negative sign the frequency of the reflected wave* ω_r.

Plasma switching in this case gives rise to the upshifted transmitted and reflected wave solutions. It will next be shown that in addition to these two traveling wave solutions, a *wiggler* mode solution exists.

Equations 3.8 through 3.10, subject to the initial conditions

$$E(0) = E_0,$$ (3.25)

$$H(0) = H_0,$$ (3.26)

$$J(0) = 0,$$ (3.27)

describe an initial value problem that can be solved easily by several techniques. The Laplace transform technique[1,2] has been chosen to lay the foundation to the solution of the more difficult transient problem of a switched plasma half-space discussed in the next chapter. By defining the Laplace transform of a function $f(t)$

$$\pounds\{f(t)\} = F(s) = \int_0^\infty f(t)\exp(-st)dt$$ (3.28)

and noting that

$$\pounds\left\{\frac{df}{dt}\right\} = sF(s) - f(0),$$ (3.29)

Equations 3.8 through 3.10 can be converted into a matrix algebraic equation with the initial conditions appearing on the right side as the excitation vector:

---

* In the literature on the subject, several alternative names are used for the waves generated by the switching action. Alternative names for the transmitted wave are (1) right-going wave, (2) positive-going wave, and (3) forward propagating wave. Alternative names for the reflected wave are (1) left-going wave, (2) negative-going wave, and (3) backward propagating wave.

$$\begin{bmatrix} \mu_0 s & -jk & 0 \\ -jk & \varepsilon_0 s & 1 \\ 0 & -\varepsilon_0 \omega_p^2 & s \end{bmatrix} \begin{bmatrix} H(s) \\ E(s) \\ J(s) \end{bmatrix} = \begin{bmatrix} \mu_0 H_0 \\ \varepsilon_0 E_0 \\ 0 \end{bmatrix}. \tag{3.30}$$

The time domain solution can then be obtained by computing the state transition matrix[1] or, more simply, by solving for the s-domain field variables and taking the Laplace inverse of each of the variables. The s-domain variables are given by

$$H(s) = H_0 \frac{s^2 + \omega_p^2 + jkcs}{s\left(s^2 + \omega_p^2 + k^2 c^2\right)}, \tag{3.31}$$

$$E(s) = E_0 \frac{s + jkc}{s^2 + \omega_p^2 + k^2 c^2}, \tag{3.32}$$

$$J(s) = \frac{H_0}{c} \frac{\omega_p^2(s + jkc)}{s\left(s^2 + \omega_p^2 + k^2 c^2\right)}. \tag{3.33}$$

It can be noted that there are two poles in the expression for $E(s)$ whereas $H(s)$ and $J(s)$ have an additional pole at the origin. The time domain solution obtained by computing the residues at the poles can be written as the sum of three modes given in Equations 3.14, 3.15, and 3.34:

$$J(t) = \sum_{m=1}^{3} J_m \exp(j\omega_m t). \tag{3.34}$$

The frequency and the fields of each mode are listed below:

Mode 1 ($m = 1$):

$$\omega_1 = \sqrt{k^2 c^2 + \omega_p^2} = \sqrt{\omega_0^2 + \omega_p^2}, \tag{3.35}$$

$$\frac{E_1}{E_0} = \frac{\omega_1 + \omega_0}{2\omega_1}, \tag{3.36}$$

$$\frac{H_1}{H_0} = \frac{\omega_0}{\omega_1} \frac{E_1}{E_0}, \tag{3.37}$$

$$J_1 = -j\varepsilon_0 \frac{\omega_p^2}{\omega_1} E_1. \tag{3.38}$$

Mode 2 ($m = 2$):

$$\omega_2 = -\sqrt{\omega_0^2 + \omega_p^2} = -\omega_1, \tag{3.39}$$

$$\frac{E_2}{E_0} = \frac{\omega_2 + \omega_0}{2\omega_2} = \frac{\omega_1 - \omega_0}{2\omega_1}, \tag{3.40}$$

$$\frac{H_2}{H_0} = \frac{\omega_0}{\omega_2} \frac{E_2}{E_o} = -\frac{\omega_0}{\omega_1} \frac{E_2}{E_0}, \tag{3.41}$$

$$J_2 = -j\varepsilon_0 \frac{\omega_p^2}{\omega_2} E_2. \tag{3.42}$$

Mode 3 ($m = 3$):

$$\omega_3 = 0, \tag{3.43}$$

$$\frac{E_3}{E_0} = 0, \tag{3.44}$$

$$\frac{H_3}{H_0} = \frac{\omega_p^2}{\omega_0^2 + \omega_p^2}, \tag{3.45}$$

$$J_3 = j\frac{\omega_0}{c} H_3, \tag{3.46}$$

$$v_3 = -\frac{1}{Nq} J_3. \tag{3.47}$$

Modes 1 and 2 are transverse electromagnetic waves whose electric and magnetic fields are related by the intrinsic impedance of the plasma medium at the frequency ω_m (ω_m is an algebraic quantity including sign):

$$\frac{E_1}{E_0} = \frac{E_0}{H_0}\frac{\omega_1}{\omega_0} = \frac{\eta_0}{\sqrt{1 - \omega_p^2/\omega_1^2}} = \frac{\eta_0}{\sqrt{\varepsilon_p(\omega_1)}} = \eta_{p1}, \tag{3.48}$$

$$\frac{E_2}{H_2} = \eta_0 \frac{\omega_2}{\omega_0} = \eta_{p2}. \tag{3.49}$$

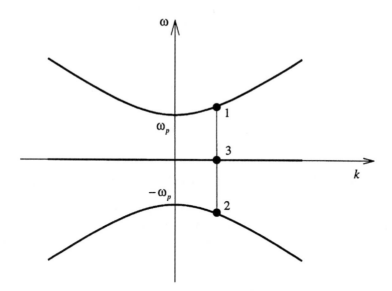

FIGURE 3.2 ω–k diagram and wiggler magnetic field.

Mode 3 is the wiggler mode discussed in the next section.[3-5] In the presence of a static magnetic field in the z-direction, the third mode becomes a traveling wave with a downshifted frequency. See Chapter 7 for a thorough discussion of this aspect.

3.3 ω–k DIAGRAM AND THE WIGGLER MAGNETIC FIELD

The ω–k diagram for the problem under discussion can be obtained from Equation 3.13 by assuming the time variation of the fields as $\exp(j\omega t)$. Under this assumption, the differentiation in time domain is equivalent to the multiplication by $(j\omega t)$ in the frequency domain:

$$(j\omega)^3 + \left[k^2 c^2 + \omega_p^2\right](j\omega) = 0$$

$$\omega\left[\omega^2 - \left(k^2 c^2 + \omega_p^2\right)\right] = 0. \tag{3.50}$$

Figure 3.2 shows the ω–k diagram where the top and the bottom branches are due to the factor in the square brackets equal to zero, and the horizontal line is due to the factor $\omega = 0$. The line $k =$ constant is a vertical line and intersects the ω–k diagram at the three points shown as 1, 2, and 3. The third mode ($\omega = 0$) is the wiggler mode. Its real fields are

$$\mathbf{E}_3(x, y, z, t) = 0, \tag{3.51}$$

$$\mathbf{H}_3(x,y,z,t) = \hat{y}H_0 \frac{\omega_p^2}{\omega_0^2 + \omega_p^2} \cos(kz), \tag{3.52}$$

$$\mathbf{J}_3(x,y,z,t) = -\hat{x}H_0 k \frac{\omega_p^2}{\omega_0^2 + \omega_p^2} \sin(kz), \tag{3.53}$$

$$\mathbf{v}_3(x,y,z,t) = -\frac{1}{Nq} \mathbf{J}_3(x,y,z,t). \tag{3.54}$$

3.4 POWER AND ENERGY CONSIDERATIONS

This section shows that, unlike in the case of spatial discontinuity, for a temporal discontinuity the real power density of the source wave is not equal to the sum of real power densities of the three modes.[6] Let $S_m = E_m H_m$. From the expressions for the fields of the various modes given above,

$$\left|\frac{S_1}{S_0}\right| + \left|\frac{S_2}{S_0}\right| + \left|\frac{S_3}{S_0}\right| = \frac{1}{2}\frac{\omega_0}{\omega_1}\left[1 + \frac{\omega_0^2}{\omega_1^2}\right] \neq 1. \tag{3.55}$$

Note that the power density of the third mode is zero.

On the other hand, it can be shown that the space-averaged energy density of the source wave is equal to the sum of the space-averaged energy densities of all the three modes. The computation of the energy density of each of the first two modes involves the stored energy in the electric and magnetic fields in free space and the kinetic energy of the electrons. The energy density of the third mode involves the energy of the wiggler magnetic field and the kinetic energy of the electrons. The steps are shown below, where (w) indicates an averaged quantity over a wavelength.

Source Wave in Free Space

$$\langle w_0 \rangle = \left\langle \tfrac{1}{2}\varepsilon_0 E_0^2 \cos^2(\omega_0 t - kz)\right\rangle + \left\langle \tfrac{1}{2}\mu_0 H_0^2 \cos^2(\omega_0 t - kz)\right\rangle = \tfrac{1}{2}\,\varepsilon_0 E_0^2. \tag{3.56}$$

Mode 1

$$\langle w_1 \rangle = \left\langle \tfrac{1}{2}\varepsilon_0 E_1^2 \cos^2(\omega_1 t - kz)\right\rangle + \left\langle \tfrac{1}{2}\mu_0 H_1^2 \cos^2(\omega_1 t - kz)\right\rangle + \tfrac{1}{2}\,Nmv_1^2. \tag{3.57}$$

From Equation 3.38 and the relation $J_1 = -qN\,v_1$, the expression for the instantaneous velocity field v_1 can be obtained:

$$v_1(x,y,z,t) = -\frac{qE_1}{m\omega_1}\sin(\omega_1 t - kz). \tag{3.58}$$

By substituting Equation 3.58 in Equation 3.57 and simplifying,

$$\frac{\langle w_1 \rangle}{\langle w_0 \rangle} = \left(\frac{E_1}{E_0} \right)^2 = \frac{1}{4} \left[1 + \frac{\omega_o}{\omega_1} \right]^2. \tag{3.59}$$

Mode 2

Similarly,

$$\frac{\langle w_2 \rangle}{\langle w_0 \rangle} = \left(\frac{E_2}{E_0} \right)^2 = \frac{1}{4} \left[1 - \frac{\omega_0}{\omega_1} \right]^2. \tag{3.60}$$

Mode 3

From Equations 3.51 and 3.52 or Equations 3.43 and 3.47,

$$w_3 = \frac{1}{2} \mu_o \left[H_3 \cos(kz) \right]^2 + \frac{1}{2} mN \left[\frac{k}{Nq} H_3 \sin(kz) \right]^2 \tag{3.61}$$

and

$$\frac{\langle w_3 \rangle}{\langle w_0 \rangle} = \frac{1}{2} \left[\frac{\omega_p^2}{\omega_0^2 + \omega_p^2} \right]^2 \left[1 + \frac{\omega_0^2}{\omega_p^2} \right] = \frac{1}{2} \frac{\omega_p^2}{\omega_0^2 + \omega_p^2}. \tag{3.62}$$

From Equations 3.59, 3.60, and 3.62,

$$\frac{\langle w_1 \rangle}{\langle w_0 \rangle} + \frac{\langle w_2 \rangle}{\langle w_0 \rangle} + \frac{\langle w_3 \rangle}{\langle w_0 \rangle} = 1. \tag{3.63}$$

3.5   PERTURBATION FROM THE STEP PROFILE*8

A step profile is a useful approximation for a fast profile with a small rise time. This section will consider a perturbation technique to compute the correction terms for a fast profile. An important step in this technique, the construction of Green's function $G(t,\tau)$ for the switched plasma medium, is considered in the next section.

---

* © 1998 IEEE. Sections 3.5 through 3.10 and Figures 3.3 through 3.10 are reprinted with permission from Huang, T. T., Lee, J. H., Kalluri, D. K., and Groves, K. M., *IEEE Transactions on Plasma Science,* 26(1), 19–25, 1998.

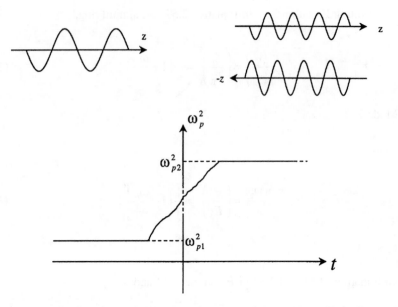

FIGURE 3.3 Perturbation from step profile. Geometry of the problem. (From Huang, T. et al., *Trans. Plasma Sci.*, 26(1), 1988. With permission.)

Figure 3.3 shows the geometry of the problem. The equation for the electric field in the ideal sudden-switching case (step profile) is given by

$$\frac{d^2 E_0(t)}{dt^2} + \tilde{\omega}^2(t)E_0(t) = 0, \tag{3.64}$$

where $\tilde{\omega}^2(t) = k^2 c^2 + \tilde{\omega}_p^2(t)$, $\tilde{\omega}_p^2(t)$ is a step function and k is the conserved wave number.[1] The solution of (1) is known:[3]

$$E_0(t) = e^{j\omega_1 t}, \qquad\qquad t < 0, \tag{3.65}$$

$$E_0(t) = T_0 e^{j\omega_1 t} R_0 e^{-j\omega_2 t}, \quad t > 0, \tag{3.66}$$

where

$$T_0 = \frac{(\omega_2 + \omega_1)}{2\omega_2}, \tag{3.67}$$

$$R_0 = \frac{(\omega_2 - \omega_1)}{2\omega_2}, \tag{3.68}$$

$$\omega_1^2 = k^2c^2 + \omega_{p1}^2, \tag{3.69}$$

$$\omega_2^2 = k^2c^2 + \omega_{p2}^2. \tag{3.70}$$

For a general profile (Figure 3.3) of $\omega_p^2(t)$, a perturbation solution is sought by using the step profile solution as a reference solution. The formulation is developed below:

$$\frac{d^2E(t)}{dt^2} + \omega^2(t)E(t) = 0, \tag{3.71}$$

$$\frac{d^2[E_0(t) + E_1(t)]}{dt^2} + [\tilde{\omega}^2(t) + \Delta\omega^2(t)][E_0(t) + E_1(t)] = 0, \tag{3.72}$$

where

$$\omega^2(t) = \omega_p^2(t) + k^2c^2, \tag{3.73}$$

$$\Delta\omega^2(t) = \omega^2(t) - \tilde{\omega}^2(t). \tag{3.74}$$

The first-order perturbation E_1 can be estimated by dropping the second-order term, $\Delta\omega^2(t)E_1(t)$, from Equation 3.72:

$$\frac{d^2E_1(t)}{dt^2} + \tilde{\omega}^2(t)E_1(t) = -\Delta\omega^2(t)E_0(t). \tag{3.75}$$

By treating the right side of Equation 3.75 as the driving function $F(t)$, the solution of E_1 can be written:

$$E_1(t) = G(t,\tau) * F(t) = \int_{-\infty}^{+\infty} G(t,\tau)\left[-\Delta\omega^2(\tau)E_0(\tau)\right]d\tau, \tag{3.76}$$

where $G(t,\tau)$ is the Green's function, which satisfies the second-order differential equation:

$$\ddot{G}(t,\tau) + \tilde{\omega}^2(t)G(t,\tau) = \delta(t-\tau). \tag{3.77}$$

The approximate one-iteration solution for the original time-varying differential equation takes the form:

$$E(t) \approx E_0(t) + E_1(t) = E_0(t) + \int_{-\infty}^{\infty} G(t,\tau)\left[-\Delta\omega^2(\tau)E_0(\tau)\right]d\tau. \tag{3.78}$$

More iterations can be made until the desired accuracy is reached. The general N-iterations solution is

$$E(t) \cong E_0(t) + \sum_{n=1}^{N} E_n(t),$$ (3.79)

where

$$E_n(t) = \int_{-\infty}^{\infty} G(t,\tau) \left[-\Delta\omega^2(\tau) E_{n-1}(\tau) \right].$$ (3.80)

3.6   CAUSAL GREEN'S FUNCTION FOR TEMPORALLY UNLIKE PLASMA MEDIA[7-9]

From the causality requirement

$$G(t,\tau) = 0, \quad t < \tau.$$ (3.81)

For $t > \tau$, $G(t, \tau)$ will be determined from the impulse source conditions,[10] i.e.,

$$G(\tau^+,\tau) = G(\tau^-,\tau),$$ (3.82)

$$\frac{\partial G(\tau^+,\tau)}{\partial t} - \frac{\partial G(\tau^-,\tau)}{\partial t} = 1.$$ (3.83)

From Equation 3.81, $G(\tau^-, \tau) = 0$ and $\partial G(\tau^-, \tau)/\partial t = 0$. Thus, Equations 3.82 and 3.83 reduce to

$$G(\tau^+,\tau) = 0,$$ (3.84)

$$\frac{\partial G(\tau^+,\tau)}{\partial t} = 1.$$ (3.85)

A geometrical interpretation of the steps involved in constructing the Green's function is given in Figure 3.4, where the horizontal axis is the spatial coordinate z along which the waves are propagating. The horizontal axis ($t = 0$) is also the temporal boundary between the two unlike plasma media.

In Figure 3.4a the case $t > 0$ is considered. The impulse source is located in the second medium. From Equation 3.77 the Green's function can be written as

$$G(t,\tau) = E_{B1}(\tau)e^{j\omega_2 t} + E_{B2}(\tau)e^{-j\omega_2 t}, \quad t > \tau > 0.$$ (3.86)

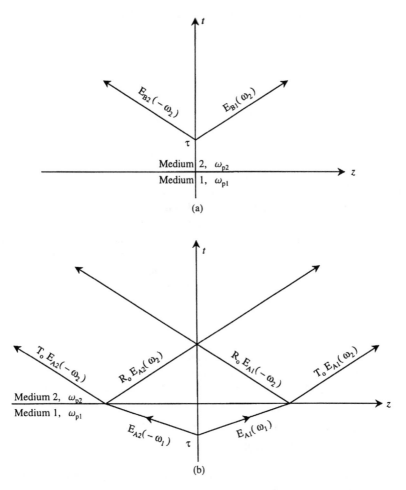

FIGURE 3.4 Geometrical interpretation of the Green's function for temporally unlike plasma media. (From Huang, T. et al., *Trans. Plasma Sci.*, 26(1), 1988. With permission.)

E_{B1} and E_{B2} are determined from Equations 3.84 and 3.85:

$$E_{B1} = \frac{1}{2j\omega_2} e^{-j\omega_2 \tau},$$ (3.87)

$$E_{B2} = -\frac{1}{2j\omega_2} e^{+j\omega_2 \tau}.$$ (3.88)

In Figure 3.4b the case $\tau < 0$ is considered. The impulse source is located in the first medium. Thus, $G(t, \tau)$, in the interval $\tau < t < 0$ can be written as

$$G(t,\tau) = E_{A1}(\tau)e^{j\omega_1 t} + E_{A2}(\tau)e^{-j\omega_1 t}, \qquad \tau < t < 0.$$ (3.89)

E_{A1} and E_{A2} are once again determined by Equations 3.84 and 3.85:

$$E_{A1} = \frac{1}{2j\omega_1} e^{-j\omega_1\tau}, \tag{3.90}$$

$$E_{A2} = -\frac{1}{2j\omega_1} e^{+j\omega_1\tau}. \tag{3.91}$$

By referring to Figure 3.4b, it can be noted that the two waves launched by the impulse source at $t = \tau$ travel along the positive z-axis (transmitted wave) and along the negative z-axis (reflected wave). They reach the temporal interface at $t = 0$, where the medium undergoes a temporal change. Each of these waves gives rise to two waves at the upshifted frequency ω_2. The amplitude of the four waves in the interval $0 < t < \infty$ can be obtained by taking into account the amplitude of each of the incident waves (incident on the temporal interface) and the scattering coefficients. The geometrical interpretation of the amplitude computation of the four waves is shown in Figure 3.4b. The Green's function for this interval $t > 0$ and $\tau < 0$ is given by

$$G(t,\tau) = E_{A1}(\tau)\left[T_0 e^{j\omega_2 t} + R_0 e^{-j\omega_2 t}\right] + E_{A2}(\tau)\left[R_0 e^{j\omega_2 t} + T_0 e^{-j\omega_2 t}\right], \quad \tau < 0 < t. \tag{3.92}$$

Explicit expressions for $G(t,\tau)$ valid for various regions of (t,τ) plane are given in Figure 3.5. The difference between this causal Green's function and the Green's function for spatially unlike dielectric media given in References 11 and 12 can be noted.

3.7 TRANSMISSION AND REFLECTION COEFFICIENTS FOR A GENERAL PROFILE[7,8]

The approximate amplitude of the electric field after one iteration ($N = 1$) can be obtained from Equation 3.76:

$$E(t) \approx E_0(t) + E_1(t)$$

$$= E_0(t) - \int_{-\infty}^{+\infty} d\tau G(t,\tau)\Delta\omega^2(\tau)\left[E_0(\tau)\right]. \tag{3.93}$$

For $t > 0$

$$E(t) \approx E_0(t) - \int_0^t d\tau \frac{\left[e^{j\omega_2(t-\tau)} - e^{j\omega_2(-t+\tau)}\right]}{2j\omega_2} \Delta\omega^2(\tau)\left[T_0 e^{j\omega_2\tau} + R_0 e^{-j\omega_2\tau}\right]$$

$$- \int_{-\infty}^0 d\tau \frac{\left[\begin{array}{c} +T_0 e^{j(\omega_2 t - \omega_1\tau)} + R_0 e^{j(-\omega_2 t - \omega_1\tau)} \\ -R_0 e^{j(\omega_2 t + \omega_1\tau)} - T_0 e^{j(-\omega_2 t + \omega_1\tau)} \end{array} \right]}{2j\omega_1} \Delta\omega^2(\tau)\left[e^{j\omega_1\tau}\right]. \tag{3.94}$$

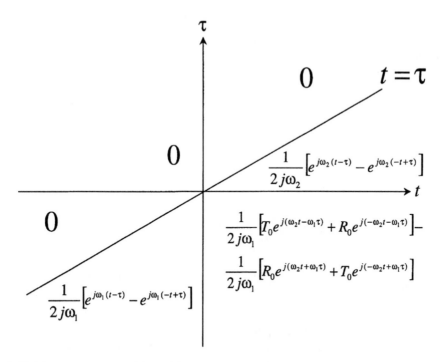

FIGURE 3.5 Causal Green's function for temporally unlike plasma media. (From Huang, T. et al., *Trans. Plasma Sci.*, 26(1), 1988. With permission.)

In the asymptotic limit of $t \to \infty$, $E_1(t)$ can be written as

$$E_1(t) = R_1 e^{-j\omega_2 t} + T_1 e^{+j\omega_2 t}, \tag{3.95}$$

where R_1 and T_1 are the first-order correction terms of the reflection coefficient and the transmission coefficient, respectively.

$$R_1 = -\int_0^{+\infty} d\tau \Delta\omega^2(\tau) \frac{-\left(R_0 + T_0 e^{j2\omega_2 \tau}\right)}{2j\omega_2} - \int_{-\infty}^0 d\tau \Delta\omega^2(\tau) \frac{\left(R_0 - T_0 e^{j2\omega_1 \tau}\right)}{2j\omega_1}, \tag{3.96}$$

$$T_1 = -\int_0^{+\infty} d\tau \Delta\omega^2(\tau) \frac{\left(T_0 + R_0 e^{j2\omega_2 \tau}\right)}{2j\omega_2} - \int_{-\infty}^0 d\tau \Delta\omega^2(\tau) \frac{\left(T_0 - R_0 e^{j2\omega_1 \tau}\right)}{2j\omega_1}. \tag{3.97}$$

Higher-order correction terms ($N > 1$) can be obtained by using more iterations.

3.8 TRANSMISSION AND REFLECTION COEFFICIENTS FOR A LINEAR PROFILE[7,8]

As a particular example, Equations 3.96 and 3.97 are used to compute R_1 and T_1 for a profile of $\omega_p^2(t)$ rising linearly from 0 (at $t = -T/2$) to ω_{p2}^2 (at $t = T/2$) with a slope of ω_{p2}^2/T_r. Here T_r is the rise time of the profile. The function $\Delta\omega^2(\tau)$ takes the form of $\omega p_2^2 (\tau/T_r - 1/2)$ for $0 < \tau < T/2$ and $\omega_{p2}^2 (\tau/T_r + 1/2)$ for $-T_r/2 < \tau < 0$. It is zero on the rest of the real line.

$$R_1 = -\int_0^{T_r/2} d\tau \omega_{p2}^2 \left(\frac{\tau}{T_r} - \frac{1}{2}\right) \frac{-\left(R_0 + T_0 e^{j2\omega_2\tau}\right)}{2j\omega_2} - \int_{-T_r/2}^0 d\tau \omega_{p2}^2 \left(\frac{\tau}{T_r} + \frac{1}{2}\right) \frac{R_0 - T_0 e^{j2\omega_1\tau}}{2j\omega_1}$$

$$= j\omega_{p2}^2 \left[-\frac{T_0\left(e^{j\omega_2 T_r} - 1 - j\omega_2 T_r\right)}{8\omega_2^3 T_r} + \frac{R_0 T_r}{16\omega_2} + \frac{T_0\left(e^{-j\omega_1 T_r} - 1 + jT_r\omega_1\right)}{8\omega_1^3 T_r} + \frac{R_0 T_r}{16\omega_1} \right],$$

(3.98)

$$T_1 = -\int_0^{T_r/2} d\tau \omega_{p2}^2 \left(\frac{\tau}{T_r} - \frac{1}{2}\right) \frac{\left(T_0 + R_0 e^{j2\omega_2\tau}\right)}{2j\omega_2} - \int_{-T_r/2}^0 d\tau \omega_{p2}^2 \left(\frac{\tau}{T_r} + \frac{1}{2}\right) \frac{\left(T_0 - R_0 e^{j2\omega_1\tau}\right)}{2j\omega_1}$$

$$= j\omega_{p2}^2 \left[+\frac{R_0\left(e^{j\omega_2 T_r} - 1 - j\omega_2 T_r\right)}{8\omega_2^3 T_r} - \frac{T_0 T_r}{16\omega_2} + \frac{R_0\left(e^{j\omega_1 T_r} - 1 + jT_r\omega_1\right)}{8\omega_1^3 T_r} + \frac{T_0 T_r}{16\omega_1} \right].$$

(3.99)

The solution for other profiles can be obtained similarly.

However, to perform the integrations in Equations 3.96 and 3.97 for an arbitrary profile, it may be necessary to expand the exponentials in Equations 3.96 and 3.97 in a power series and then do the integration. For a given T_r, it is possible to choose the relative positioning of $\omega_p^2(t)$ and $\tilde{\omega}_p^2(t)$, such that

$$\int_{-\infty}^{+\infty} \Delta\omega^2(\tau)d\tau = 0. \tag{3.100}$$

Such a choice will improve the accuracy of the power reflection coefficient if one chooses to approximate the exponential terms in Equations 3.98 and 3.99 by keeping only one term (the dominant term). For the linear profile under consideration, Equation 3.100 is satisfied, and by keeping only the dominant term, the following is obtained:

$$R \approx R_0 + R_1 = R_0 - \frac{\omega_{p2}^2 T_r^2}{24} T_0, \tag{3.101}$$

$$T \approx T_0 + T_1 = T_0 - \frac{\omega_{p2}^2 T_r^2}{24} R_0. \tag{3.102}$$

3.9 VALIDATION OF THE PERTURBATION SOLUTION BY COMPARING WITH THE EXACT SOLUTION[7,8]

To illustrate the validity of the method, the perturbation solution using Equations 3.98 and 3.99 is compared with the exact solution of a linear profile. For a linear profile, Equation 3.71 can be arranged into the form of the Airy differential equation by introducing a new variable ξ. The basic steps are shown below:

$$\xi = \omega^2(t) = k^2 c^2 + \omega_{p1}^2 + \frac{\omega_{p2}^2 - \omega_{p1}^2}{T_r} t, \qquad 0 < t < T_r, \qquad (3.103)$$

$$\frac{d\xi}{dt} = \frac{\omega_{p2}^2 - \omega_{p1}^2}{T_r} = \beta, \qquad (3.104)$$

$$\frac{d^2 E}{dt^2} + \omega^2(t) E = 0, \qquad (3.105)$$

$$\frac{d^2 E}{d\xi^2} \beta^2 + \xi E = 0, \qquad (3.106)$$

$$\frac{d^2 E}{d\xi^2} + \frac{\xi}{\beta^2} E = 0. \qquad (3.107)$$

The solution of Equation 3.107 is given by

$$E(t) = \begin{cases} e^{j\omega_1 t}, & t < 0 \\ c_1 Ai\left[-\beta^{-2/3}\xi\right] + c_2 Bi\left[-\beta^{-2/3}\xi\right], & 0 < t < T_r \\ Te^{j\omega_2 t} + Re^{-j\omega_2 t}, & t > T_r. \end{cases} \qquad (3.108)$$

where Ai and Bi are the Airy functions.[13]

The four coefficients c_1, c_2, T, and R in Equation 3.108 can be determined by using the continuity conditions on E and $\dot{E}$ at the two temporal interfaces $t = 0$ and $t = T_r$. Thus, the scattering coefficients R and T are obtained but are not given here to save space. It can be noted that the power reflection coefficient computed from the exact solution depends on T_r, but is independent of the location of the starting point of the linear profile.

Figures 3.6 and 3.7 compare the perturbation solution (broken line) based on Equations 3.98 and 3.99 with the exact solution (solid line) for the power reflection coefficient ($\rho = |R|^2 (\omega_1/\omega_2)$) and power transmission coefficient ($\tau = |T|^2 (\omega_1/\omega_2)$) for a linear profile. The results are presented in a normalized form by taking $f_1 = \omega_1/2\pi = 1.0$. The other parameters are $f_{p1} = \omega_{p1}/2\pi = 0$ and $f_{p2} = \omega_{p2}/2\pi = 1.2$. The perturbation solution tracks the exact solution closely up to $T_r = 0.25$ (at this point the power reflection coefficient is ¼ of the value for sudden switching). More iterations can be taken by increasing N in Equation 3.80 to extend the range of validity of the perturbation solution. It can be noted that both ρ and τ decrease as T_r increases. A reduction in ρ does not result in an increase in τ. At a temporal discontinuity, $\rho + \tau \neq 1$.[6,14]

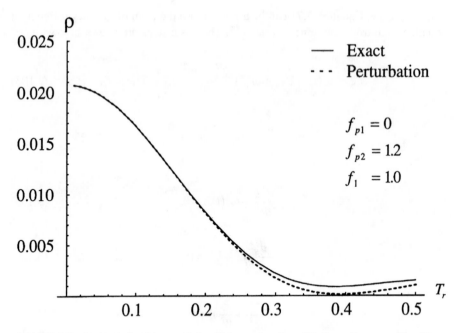

FIGURE 3.6   Power reflection coefficient (ρ) vs. rise time (T_r) for a linear profile. (From Huang, T. et al., *Trans. Plasma Sci.*, 26(1), 1988. With permission.)

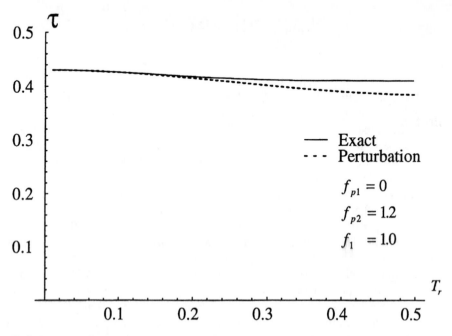

FIGURE 3.7   Power transmission coefficient (τ) vs. rise time (T_r) for a linear profile. (From Huang, T. et al., *Trans. Plasma Sci.*, 26(1), 1988. With permission.)

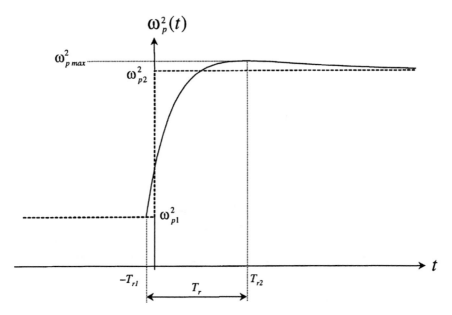

FIGURE 3.8   Sketch of a hump profile. (From Huang, T. et al., *Trans. Plasma Sci.*, 26(1), 1988. With permission.)

3.10 HUMP PROFILE[8]

For a general profile, which has no exact solution, Equations 3.96 and 3.97 provide a systematic way to compute the scattering coefficients. If a transient plasma with a "hump" profile is chosen to illustrate the general applicability, the electron density increases rapidly and then settles gradually to a new level. The profile is given by

$$\omega_p^2(t) = \omega_{p1}^2, \qquad\qquad\qquad -\infty < t < -T_{r1}, \qquad (3.109)$$

$$\omega_p^2(t) = \omega_{p1}^2 e^{-a(t+T_{r1})} + \omega_{p2}^2\left[1 - e^{-b(t+T_{r1})}\right], \quad -T_{r1} < t < \infty. \qquad (3.110)$$

The parameter T_{r1} is chosen such that Equation 3.100 is satisfied. Thus,

$$T_{r1} = \frac{\dfrac{\omega_{p2}^2}{b} - \dfrac{\omega_{p1}^2}{a}}{\omega_{p2}^2 - \omega_{p1}^2}. \qquad (3.111)$$

Figure 3.8 shows the profile. The profile has an extremum at $t = T_{r2}$, where

$$T_{r2} = \frac{1}{b-a}\left[\ln\frac{b}{a} + \ln\frac{\omega_{p2}^2}{\omega_{p1}^2}\right] - T_{r1}. \qquad (3.112)$$

The rise time T_r is defined as

$$T_r = T_{r1} + T_{r2} = \frac{1}{b-a}\left[\ln\frac{b}{a} + \ln\frac{\omega_{p2}^2}{\omega_{p1}^2}\right]. \tag{3.113}$$

In the above, we restrict the two parameters a and b by the inequality

$$1 < \frac{b}{a} < \frac{\omega_{p2}^2}{\omega_{p1}^2}, \tag{3.114}$$

which ensures that T_{r1} is positive and the extremum is a maximum.

Equation 3.71 has no obvious exact solution for the hump profile. However, with the perturbation approach of Equations 3.96 and 3.97, it is possible to solve for the transmission and reflection coefficients explicitly. For the hump profile,

$$\Delta\omega^2(t) = \begin{cases} 0, & -\infty < t < -T_{r1} \\ \omega_{p1}^2 e^{-a(t+T_{r1})} + \omega_{p2}^2\left[1 - e^{-b(t+T_{r1})}\right] - \omega_{p1}^2, & -T_{r1} < t < 0 \\ \omega_{p1}^2 e^{-a(t+T_{r1})} + \omega_{p2}^2\left[1 - e^{-b(t+T_{r1})}\right] - \omega_{p2}^2, & 0 < t < \infty, \end{cases} \tag{3.115}$$

and the transmission and reflection coefficients, based on one iteration, are

$$R \approx R_0 - \int_0^{+\infty} d\tau \left[\omega_{p1}^2 e^{-a(\tau+T_{r1})} - \omega_{p2}^2 e^{-b(\tau+T_{r1})}\right]\frac{\left(R_0 + T_0 e^{2j\omega_2\tau}\right)}{-2j\omega_2}$$

$$-\int_{-T_{r1}}^0 d\tau \left[\omega_{p1}^2\left(e^{-a(\tau+T_{r1})} - 1\right) - \omega_{p2}^2\left(e^{-b(\tau+T_{r1})} - 1\right)\right]\frac{\left(R_0 - R_0 e^{2j\omega_1\tau}\right)}{2j\omega_1}$$

$$= R_0 + \frac{R_0}{2j\omega_2}\left(\frac{\omega_{p1}^2 e^{-aT_{r1}}}{a} - \frac{\omega_{p2}^2 e^{-bT_{r1}}}{b}\right) + \frac{T_0}{2j\omega_2}\left(\frac{\omega_{p1}^2 e^{-aT_{r1}}}{a - 2j\omega_2} - \frac{\omega_{p2}^2 e^{-bT_{r1}}}{b - 2j\omega_2}\right)$$

$$+ \frac{T_0}{2j\omega_1}\left(\left(\omega_{p2}^2 - \omega_{p1}^2\right)\frac{1 - e^{-2j\omega_1 T_{r1}}}{2j\omega_1} - \omega_{p1}^2\frac{e^{-aT_{r1}} - e^{-2j\omega_1 T_{r1}}}{a - 2j\omega_1} + \omega_{p2}^2\frac{e^{-bT_{r1}} - e^{-2j\omega_1 T_{r1}}}{b - 2j\omega_1}\right)$$

$$- \frac{R_0}{2j\omega_1}\left(\frac{\omega_{p1}^2}{a}\left(1 - e^{-aT_{r1}} - aT_{r1}\right) - \frac{\omega_{p1}^2}{b}\left(1 - e^{-bT_{r1}} - bT_{r1}\right)\right) \tag{3.116}$$

$$T \approx T_0 - \int_0^{+\infty} d\tau \left[\omega_{p1}^2 e^{-a(\tau+T_{r1})} - \omega_{p2}^2 e^{-b(\tau+T_{r1})} \right] \frac{\left(T_0 + R_0 e^{-2j\omega_2\tau} \right)}{2j\omega_2}$$

$$- \int_{-T_{r1}}^0 d\tau \left[\omega_{p1}^2 \left(e^{-a(\tau+T_{r1})} - 1 \right) - \omega_{p2}^2 \left(e^{-b(\tau+T_{r1})} - 1 \right) \right] \frac{\left(T_0 - R_0 e^{2j\omega_1\tau} \right)}{2j\omega_1}$$

$$= T_0 - \frac{T_0}{2j\omega_2} \left(\frac{\omega_{p1}^2 e^{-aT_{r1}}}{a} - \frac{\omega_{p2}^2 e^{-bT_{r1}}}{b} \right) - \frac{R_0}{2j\omega_2} \left(\frac{\omega_{p1}^2 e^{-aT_{r1}}}{a+2j\omega_2} - \frac{\omega_{p2}^2 e^{-bT_{r1}}}{b+2j\omega_2} \right)$$

$$+ \frac{R_0}{2j\omega_1} \left(\left(\omega_{p2}^2 - \omega_{p1}^2 \right) \frac{1 - e^{-2j\omega_1 T_{r1}}}{2j\omega_1} - \omega_{p1}^2 \frac{e^{-aT_{r1}} - e^{-2j\omega_1 T_{r1}}}{a-2j\omega_1} + \omega_{p2}^2 \frac{e^{-bT_{r1}} - e^{-2j\omega_1 T_{r1}}}{b-2j\omega_1} \right)$$

$$- \frac{T_0}{2j\omega_1} \left(\frac{\omega_{p1}^2}{a} \left(1 - e^{-aT_{r1}} - aT_{r1} \right) - \frac{\omega_{p1}^2}{b} \left(1 - e^{-bT_{r1}} - bT_{r1} \right) \right). \tag{3.117}$$

By using Equations 3.116 and 3.117, the power reflection and transmission coefficients are computed and presented in Figures 3.9 and 3.10. The parameters are $f_{p1} = 0.6$, $f_{p2} = 1.2$, $f_1 = 1.0$, and $b/a = 2.5$. The value of T_r is changed from 0^+ to 0.5 by varying the individual values of a and b. The choice of a constant value for the ratio b/a results in the same $\omega_{p\,max}^2$ for various profiles with different T_r. The unexpected slight increase of ρ at the start of the curve is worth noting. When T_r is 0^+, the effect of the hump is not felt and the scattering coefficients are those of the reference profile. As T_r increases, the effect of the hump is a rapid change of ω_p from ω_{p1} to $\omega_{p\,max}$ and the power scattering coefficients are close to the values expected for sudden switching from ω_{p1} to $\omega_{p\,max}$ rather than ω_{p1} to ω_{p2}. As T_r increases further, the rapid change is no longer close to the sudden-switching approximation. R_1 and T_1 have significant values and result in the decrease of the power scattering coefficients. By adjusting the four parameters of the hump profile, many practical transient plasma profiles can be studied.

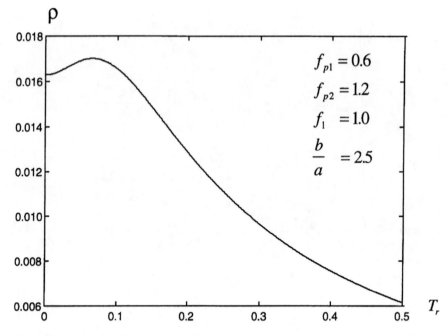

FIGURE 3.9   Power reflection coefficient (ρ) vs. rise time ($T_r = T_{r1} + T_{r2}$) for a hump profile. (From Huang, T. et al., *Trans. Plasma Sci.*, 26(1), 1988. With permission.)

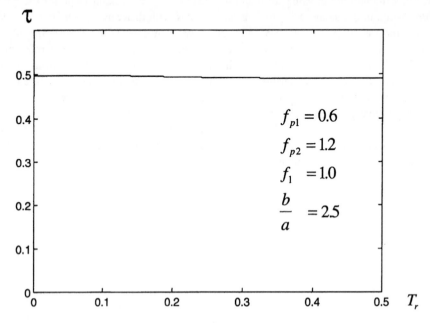

FIGURE 3.10   Power transmission coefficient (τ) vs. rise time (T_r) for a hump profile. (From Huang, T. et al., *Trans. Plasma Sci.*, 26(1), 1988. With permission.)

REFERENCES

1. Derusso, P. M., Roy, R. J., and Close, C. H., *State Variables for Engineers*, John Wiley & Sons, New York, 1965.
2. Aseltine, J. A., *Transform Methods in Linear System Analysis*, McGraw-Hill, New York, 1958.
3. Jiang, C. L., Wave propagation and dipole radiation in a suddenly created plasma, *IEEE Trans. Antennas Propag.*, AP-23, 83, 1975.
4. Kalluri, D. K., On reflection from a suddenly created plasma half-space: transient solution, *IEEE Trans. Plasma Sci.*, 16, 11, 1988.
5. Wilks, S. C., Dawson, J. M., and Mori, W. B., Frequency up-conversion of electromagnetic radiation with use of an overdense plasma, *Phys. Rev. Lett.*, 61, 337, 1988.
6. Auld, B. A., Collins, J. H., and Zapp, H. R., Signal processing in a nonperiodically time-varying magnetoelastic medium, *Proc. IEEE*, 56, 258, 1968.
7. Kalluri, D. K., Green's function for a switched plasma medium and a perturbation technique for the study of wave propagation in a transient plasma with a small rise time, in *Conf. Rec. Abstracts, IEEE Int. Conf. Plasma Science,* Boston, MA, 1996.
8. Huang, T. T., Lee, J. H., Kalluri, D. K., and Groves, K. M., Wave propagation in a transient plasma: development of a Green's function, *IEEE Trans. Plasma Sci.*, 1, 19, 1998.
9. Felsen, L. B. and Whitman, G. M., Wave propagation in time-varying media, *IEEE Trans. Antennas Propag.*, AP-18, 242, 1970.
10. Ishimaru, A., *Electromagnetic Wave Propagation, Radiation, and Scattering*, Prentice-Hall, Englewood Cliffs, NJ, 1991.
11. Lekner, J., *Theory of Reflection*, Kluwer, Boston, 1987.
12. Triezenberg, D. G., Capillary Waves in a Diffuse Liquid-Gas Interface, Ph.D. Thesis, University of Maryland, College Park, 1973.
13. Abramowitz, M. and Stegun, I. A., *Handbook of Mathematical Functions*, Dover Publications, New York, 1965.
14. Kalluri, D. K., Frequency upshifting with power intensification of a whistler wave by a collapsing plasma medium, *J. Appl. Phys.*, 79, 3895, 1996.

4 Switched Plasma Half-Space: *A* and *B* Waves*

4.1 INTRODUCTION

Chapter 2 considered the incidence of a source wave on a spatial step discontinuity in the dielectric properties of a time-invariant medium. The resulting reflected and transmitted waves have the same frequency as the source wave, and these will be labeled *A* waves. Chapter 3 considered the incidence of a source wave on a temporal step discontinuity in the dielectric properties of a space-invariant plasma medium. The resulting reflected and transmitted waves have an upshifted frequency,[1-4] and these will be labeled *B* waves. Step profiles are mathematical approximations for fast profiles and provide insight into the physical processes as well as serve as *reference solutions* for a perturbation technique (see Sections 3.5 through 3.10).

This chapter considers a problem that involves simultaneous consideration of the effects of a temporal discontinuity and a spatial discontinuity.

4.2 STEADY-STATE SOLUTION

Figure 4.1 shows the electron density $N(t)$ of a typical transient plasma and its approximation by a step profile. Figure 4.2 shows the geometry of the problem. A plane wave of frequency ω_0 is traveling in free space in the z-direction when, at $t = 0$, a semi-infinite plasma of electron density N_0 is created in the upper half of the plane $z > 0$. The reflected wave will have two components denoted by the subscripts *A* and *B*. There is no special significance to the choice of these letters for the subscripts.

The *A* component is due to reflection at the spatial discontinuity at $z = 0$, and the subscript *S* will be used to indicate scattering at a spatial boundary. The corresponding reflection coefficient is

$$R_A = R_S = \frac{\eta_{p0} - \eta_0}{\eta_{p0} + \eta_0},$$ (4.1)

* © IEEE. Reprinted with permission from Kalluri, D. K., *IEEE Transactions on Plasma Science*, 16, 11–16, 1988. Chapter 4 is an adaptation of the reprint.

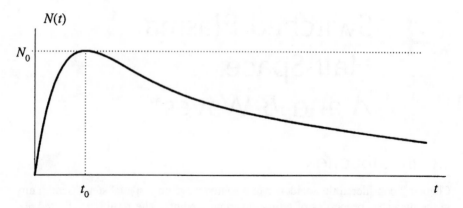

FIGURE 4.1   Electron density $N(t)$ vs. time of a typical transient plasma. (From Kalluri, D. K., *IEEE Trans. Plasma Sci.,* 16, 11–16, 1988. With permission.)

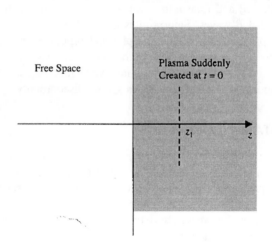

FIGURE 4.2   Geometry of the steady-state problem. (From Kalluri, D. K., *IEEE Trans. Plasma Sci.,* 16, 11–16, 1988. With permission.)

where

$$\eta_{p0} = \frac{\eta_0}{\sqrt{1 - \omega_p^2/\omega_0^2}} \tag{4.2}$$

and

$$\omega_p = \sqrt{\frac{N_0 q^2}{m\varepsilon_0}}. \tag{4.3}$$

Here, η_0 is the characteristic impedance of free space, η_{p0} is the characteristic impedance of the plasma medium, and the other symbols have the usual meanings.[5]

The B component arises as a result of the wave reflected by the temporal discontinuity at $t = 0$ and, say, $z = z_1$. Jiang[3] showed that this wave is of a new frequency ω,

$$\omega = \sqrt{\omega_0^2 + \omega_p^2},\qquad (4.4)$$

and has a relative amplitude (relative to the incident wave amplitude)

$$R_t = \frac{\omega - \omega_0}{2\omega}.\qquad (4.5)$$

The subscript t indicates scattering due to the temporal discontinuity. The wave will travel along the negative z-axis. When it reaches the spatial boundary at $z = 0$, a part of it will be transmitted into free space. The corresponding transmission coefficient is

$$T_S = \frac{2\eta_0}{\eta_p + \eta_0},\qquad (4.6)$$

where

$$\eta_p = \frac{\eta_0}{\sqrt{1 - \omega_p^2/\omega^2}}.\qquad (4.7)$$

The B component of the reflected wave in free space has a relative amplitude

$$R_B = R_t T_S.\qquad (4.8)$$

In considering the steady-state values ($t \to \infty$), the B component can be ignored since it will be damped out in traveling from $z = z_1 = \infty$ to $z = 0$, even if the plasma is only slightly lossy.

A quantitative idea of the damping of the B wave will now be given by calculating the damping constants for a low-loss plasma. Let ν be the collision frequency, and let $(\nu/\omega_p) \ll 1$. The attenuation constant α of the B wave of frequency ω is given by Reference 5, pp. 7–9:

$$\alpha = \left(\frac{\omega}{c}\right)\left(\frac{\omega_p^2}{2\omega^3}\right)\bigg/\sqrt{1 - \frac{\omega_p^2}{\omega^2}}, \quad \nu \ll \omega_p.\qquad (4.9)$$

After simplifying the algebra,

$$\alpha = \frac{\nu\omega_p^2}{2\omega_0\omega c}, \quad \nu \ll \omega_p.\qquad (4.10)$$

Equation 4.10 shows that α is large for small values of ω_0. When the incident wave frequency ω_0 is low, the frequency ω of the B wave is only slightly larger than the plasma frequency ω_p and the B wave is heavily attenuated.

Because of attenuation, R_B is now modified as

$$R_B = R_t T_s \exp(-\alpha z_1).$$

(4.11)

A damping distance constant z_p can now be defined:

$$z_p = \frac{1}{\alpha}.$$

(4.12)

In traveling this distance z_p, the B wave attenuates to e^{-1} of its original value. The attenuation can be expressed in terms of time rather than distance by noting

$$t_p = \frac{z_p}{v_g},$$

(4.13)

where t_p is the damping time constant of the B wave and v_g is the group velocity of propagation of the B wave given by

$$v_g = c\sqrt{1 - \frac{\omega_p^2}{\omega^2}} = c\frac{\omega_0}{\omega}.$$

(4.14)

From Equations 4.10 and 4.12 through 4.14,

$$t_p = \left(\frac{2\omega^2}{\nu\omega_p^2}\right), \quad \nu \ll \omega_p.$$

(4.15)

The amplitude of the B wave reduces to e^{-1} of its original value in a time t_p.

Numerical results are discussed in terms of normalized values. Let

$$\Omega_c = \frac{\nu}{\omega_p},$$

(4.16)

$$\Omega_0 = \frac{\omega_0}{\omega_p}.$$

(4.17)

In terms of these variables, from Equations 4.1, 4.5, 4.6, 4.12, and 4.13, Equations 4.18 through 4.23 are obtained:

$$R_S = 1, \qquad \Omega_0 < 1, \tag{4.18}$$

$$R_S = \frac{\Omega_0 - \sqrt{\Omega_0^2 - 1}}{\Omega_0 + \sqrt{\Omega_0^2 - 1}}, \qquad \Omega_0 > 1, \tag{4.19}$$

$$R_t = \frac{\sqrt{\Omega_0^2 + 1} - \Omega_0}{2\sqrt{\Omega_0^2 + 1}}, \tag{4.20}$$

$$T_S = \frac{2\Omega_0}{\Omega_0 + \sqrt{\Omega_0^2 + 1}}, \tag{4.21}$$

$$\omega_p t_p = \frac{2(\Omega_0^2 + 1)}{\Omega_c}, \tag{4.22}$$

$$\frac{z_p}{\lambda_p} = \frac{\Omega_0 \sqrt{\Omega_0^2 + 1}}{\pi \Omega_c}, \tag{4.23}$$

where λ_p is the free-space wavelength corresponding to the plasma frequency:

$$\lambda_p = \frac{2\pi c}{\omega_p}. \tag{4.24}$$

Figure 4.3 shows $R_A = R_S$, R_t, T_S, $R_B = R_t T_S$ vs. Ω_0. The graph of R_t shows that the amplitude of the reflected wave generated in the plasma by the temporal discontinuity decreases as the incident wave frequency ω_0 increases (plasma frequency ω_p being kept constant). This is to be expected since the plasma behaves like free space when $\omega_0 \gg \omega_p$. On the other hand, the transmission coefficient T_S at the spatial discontinuity increases with ω_0. Thus, the reflection coefficient of the B wave increases, attains a maximum, and then decreases as ω_0 is increased. The variation of the two reflection coefficients (R_A and R_B) is shown in the figure and discussed next.

For $\Omega_0 < 1$, $|R_A|$ is 1 and R_B reaches a peak value of about 17% at about $\Omega_0 = 0.5$. For $\Omega_0 > 1$, R_A falls quickly and R_B slowly, both reaching about 5% at $\Omega_0 = 2$. For $\Omega_0 > 2$, $R_B/R_A \approx 1$. When the incident wave frequency is large compared with the plasma frequency, the total reflection coefficient is low, but each component contributes significantly to this low value.

Figure 4.4 shows the damping time constant t_p for $\Omega_c = 0.01$ (log–log scale used). From this graph it is evident that for large ω_0, the B component persists for a long time in low-loss plasma. For small ω_0, this component is damped out rather quickly. Figure 4.5 shows the damping distance constant z_p.

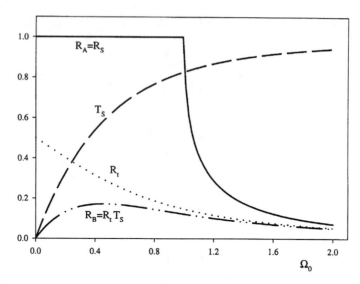

FIGURE 4.3 Reflection coefficients vs. Ω_0 (normalized incident wave frequency). (From Kalluri, D. K., *IEEE Trans. Plasma Sci.,* 16, 11–16, 1988. With permission.)

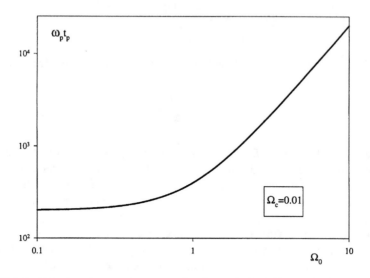

FIGURE 4.4 Damping time constant t_p vs. Ω_0. (From Kalluri, D. K., *IEEE Trans. Plasma Sci.,* 16, 11–16, 1988. With permission.)

4.3 TRANSIENT SOLUTION

4.3.1 FORMULATION AND SOLUTION

Figure 4.6 shows the geometry of the problem. A perpendicularly polarized plane wave is propagating in the z-direction when, at $t = 0$, a semi-infinite plasma of particle density N_0 is created in the upper half of the z-plane.

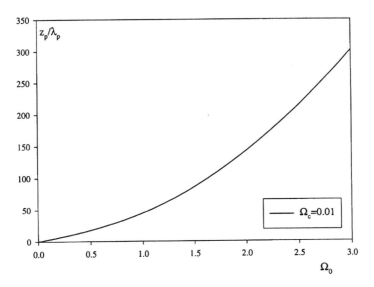

FIGURE 4.5 Damping distance constant z_p vs. Ω_0. (From Kalluri, D. K., *IEEE Trans. Plasma Sci.,* 16, 11–16, 1988. With permission.)

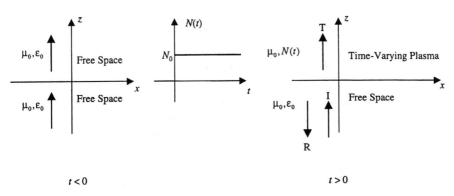

FIGURE 4.6 Geometry of the transient problem. (From Kalluri, D. K., *IEEE Trans. Plasma Sci.,* 16, 11–16, 1988. With permission.)

Let the electric field of the incident wave be

$$\mathbf{E}(\mathbf{r},t) = E_y(z,t)\hat{y}, \qquad t < 0, -\infty < z < \infty, \tag{4.25}$$

$$E_y(z,t) = E_0 \cos(\omega_0 t - k_0 z) = \mathrm{Re}\left[E_0 \exp\left\{j(\omega_0 t - k_0 z)\right\}\right], \tag{4.26}$$

and $k_0 = \omega_0/c$. Hereafter, Re will be omitted and understood. The equations satisfied by E_y are

$$\frac{\partial^2 E_{y1}}{\partial z^2} - \frac{1}{c^2}\frac{\partial^2 E_{y1}}{\partial t^2} = 0, \qquad t > 0, z < 0, \tag{4.27}$$

and from Equation 1.24

$$\frac{\partial^2 E_{y2}}{\partial z^2} - \frac{1}{c^2}\frac{\partial^2 E_{y2}}{\partial t^2} - \frac{\omega_p^2}{c^2}E_{y2} = 0, \qquad t > 0, z > 0. \tag{4.28}$$

Here, ω_p is the plasma frequency defined in Equation 4.3. The subscript 1 indicates fields in the lower half of the x–z plane ($z < 0$), and the subscript 2 indicates fields in the upper half ($z > 0$). Let the Laplace transform of $E_y(z,t)$ be $E_y(z,s)$:

$$\mathcal{L}\{E_y(z,t)\} = E_y(z,s). \tag{4.29}$$

Equation 4.28 is transformed into (here, $' = \partial/\partial t$)

$$\frac{d^2 E_{y2}}{dz^2} - \frac{1}{c^2}\left(s^2 + \omega_p^2\right)E_{y2} + \frac{s}{c^2}E_{y2}(z,0) + \frac{1}{c^2}E_{y2}'(z,0) = 0. \tag{4.30}$$

From the initial conditions,

$$E_{y1}(z,0) = E_{y2}(z,0) = E_0 \exp\left(-jk_0 z\right) \quad \text{and} \tag{4.31}$$

$$E_{y1}'(z,0) = E_{y2}'(z,0) = j\omega_0 E_0 \exp\left(-jk_0 z\right). \tag{4.32}$$

In the above, it is assumed that at $t = 0^+$ the newly created electrons and ions of the plasma are stationary so that the tangential components of the electric and magnetic fields are the same at $t = 0^-$ as at $t = 0^{+}$.[3] From Equations 4.30 through 4.32,

$$\left[\frac{d^2}{dz^2} + \frac{s^2 + \omega_p^2}{c^2}\right]E_{y2}(z,s) = -\left[\frac{s + j\omega_0}{c^2}\right]E_0 \exp\left(-jk_0 z\right). \tag{4.33}$$

The solution of this ordinary differential equation is given by

$$E_{y2}(z,s) = A_2(s)\exp\left(-q_2 z\right) + \left[\frac{s + j\omega_0}{s^2 + \omega_0^2 + \omega_p^2}\right]E_0 \exp\left(-jk_0 z\right), \tag{4.34}$$

where

$$q_2 = \sqrt{s^2 + \omega_p^2}\Big/c. \tag{4.35}$$

Similarly, from Equations 4.27, 4.31, and 4.32,

$$E_{y1}(z,s) = A_1(s)\exp(q_1 z) + \left[\frac{s+j\omega_0}{s^2+\omega_0^2}\right]E_0 \exp(-jk_0 z), \qquad (4.36)$$

where

$$q_1 = s/c. \qquad (4.37)$$

The second term on the right side of Equation 4.36 is the Laplace transform of the incident electric field. Therefore, the first term on the right side of Equation 4.36 is due to the reflected electric field. The undetermined constant $A_1(s)$ and $A_2(s)$ can be obtained from the boundary conditions of continuity of the tangential E and H, Equations 4.38 and 4.39:

$$E_{y1}(0,s) = E_{y2}(0,s), \qquad (4.38)$$

$$\left[\frac{\partial E_{y1}(z,s)}{\partial z}\right]_{z=0} = \left[\frac{\partial E_{y2}(z,s)}{\partial z}\right]_{z=0}, \qquad (4.39)$$

$$\frac{A_1(s)}{E_0} = \left[\frac{j\omega_0 - \sqrt{s^2+\omega_p^2}}{s+\sqrt{s^2+\omega_p^2}}\right]\left[\frac{(s+j\omega_0)\omega_p^2}{(s^2+\omega_0^2+\omega_p^2)(s^2+\omega_0^2)}\right]. \qquad (4.40)$$

The reflected electric field is given by

$$E_{yr}(z,t)/E_0 = \pounds^{-1}\left[\{A_{1R}(s)/E_0\}\{\exp(q_1 z)\}\right], \qquad (4.41)$$

where $A_{1R}(s)$ is the real part of $A_1(s)$:

$$\left[\frac{A_{1R}(s)}{E_0}\right] = \left[\frac{\omega_p^2\left(-\omega_0^2 - s\sqrt{s^2+\omega_p^2}\right)}{(s^2+\omega_0^2)(s^2+\omega_0^2+\omega_p^2)(s+\sqrt{s^2+\omega_p^2})}\right]. \qquad (4.42)$$

After some algebraic manipulation, Equation 4.42 can be written as the sum of five terms:

$$\text{TERM1} = \frac{2\omega_0^2-\omega_p^2}{\omega_p^2}\frac{s}{s^2+\omega_0^2}, \qquad (4.43)$$

$$\text{TERM2} = \frac{2\omega_0^2\left(\omega_0^2-\omega_p^2\right)}{\omega_p^2}\frac{1}{\sqrt{s^2+\omega_p^2}\left(s^2+\omega_0^2\right)}, \qquad (4.44)$$

$$\text{TERM3} = \frac{1}{\sqrt{s^2 + \omega_p^2}}, \tag{4.45}$$

$$\text{TERM4} = \frac{-2\omega_0^2}{\omega_p^2} \frac{s}{s^2 + \omega_0^2 + \omega_p^2}, \tag{4.46}$$

$$\text{TERM5} = \frac{-\left(2\omega_0^2 + \omega_p^2\right)\omega_0^2}{\omega_p^2} \frac{1}{\sqrt{s^2 + \omega_p^2}\left(s^2 + \omega_0^2 + \omega_p^2\right)}. \tag{4.47}$$

The Laplace inverses of Equations 4.43, 4.45, and 4.46 are known: these are
term1, term3, and term4 given in Equations 4.50, 4.52, and 4.53, respectively, but
the inversion of Equations 4.44 and 4.47 required further work. The author has
developed an algorithm[7] for numerical Laplace inversion and computation of a class
of Bessel-like functions of the type

$$h_{mn}(a,b,t) = \pounds^{-1} 1 \Big/ \Big[\left(s^2 + a^2\right)^{(n+1/2)}\left(s^2 + b^2\right)^m\Big]. \tag{4.48}$$

In terms of this notation and from the well-known Laplace transform pairs,[8]

$$E_{yr}(0,t)/E_0 = \text{term1} + \text{term2} + \text{term3} + \text{term4} + \text{term5}, \tag{4.49}$$

where

$$\text{term1} = \frac{2\omega_0^2 - \omega_p^2}{\omega_p^2} \cos\omega_0 t, \tag{4.50}$$

$$\text{term2} = \frac{2\omega_0^2\left(\omega_0^2 - \omega_p^2\right)}{\omega_p^2} h_{10}\left(\omega_p, \omega_0, t\right), \tag{4.51}$$

$$\text{term3} = J_0\left(\omega_p t\right), \tag{4.52}$$

$$\text{term4} = -\frac{2\omega_0^2}{\omega_p^2} \cos\left(\sqrt{\omega_0^2 + \omega_p^2}\, t\right), \tag{4.53}$$

$$\text{term5} = \frac{-\left(2\omega_0^2 + \omega_p^2\right)\omega_0^2}{\omega_p^2} h_{10}\left(\omega_p, \sqrt{\omega_0^2 + \omega_p^2}, t\right). \tag{4.54}$$

Here, $E_{yr}(z,t)/E_0$ is obtained by replacing t with $(t+z/c)$ in Equations 4.50 through 4.54.

4.3.2 Steady-state Solution from the Transient Solution

The steady-state value of $h_{10}(a,b,t)$ can be shown to be[7]

$$\lim_{t\to\infty} h_{10}(a,b,t) = \frac{1}{b\sqrt{a^2-b^2}}\sin(bt), \qquad b < a, \tag{4.55}$$

$$\lim_{t\to\infty} h_{10}(a,b,t) = -\frac{1}{b\sqrt{b^2-a^2}}\cos(bt), \qquad b > a. \tag{4.56}$$

From Equations 4.55 and 4.56 it can be shown that E_{LA}, the limiting value (value as t tends to infinity) of {term1 + term2} in Equation 4.49, gives Equations 4.57 and 4.58:

$$E_{LA} = \left[2\frac{\omega_0^2-\omega_p^2}{\omega_p^2}\right]\cos(\omega_0 t) - \left[2\omega_0\frac{\sqrt{\omega_p^2-\omega_0^2}}{\omega_p^2}\right]\sin(\omega_0 t), \quad \omega_0 < \omega_p, \tag{4.57}$$

$$E_{LA} = \left[2\omega_0^2-\omega_p^2-2\omega_0\frac{\sqrt{\omega_0^2-\omega_p^2}}{\omega_p^2}\right]\cos(\omega_0 t), \qquad \omega_0 > \omega_p. \tag{4.58}$$

The steady-state solution $t \to \infty$, in the presence of the space boundary alone, is easily shown to be

$$\left.\frac{E_{yr}(0,t)}{E_0}\right|_{\substack{\text{steady-state}\\\text{space boundary}}} = \mathrm{Re}\left[\left\{\frac{j\omega_0-\sqrt{\omega_p^2-\omega_0^2}}{j\omega_0+\sqrt{\omega_p^2-\omega_0^2}}\right\}\exp(j\omega_0 t)\right] \tag{4.59}$$

The term in the curly brackets, { }, is the reflection coefficient R_A of Equation 4.2. The right sides of Equations 4.57 through 4.59 are the same, and, hence, E_{LA} gives the steady-state value of the reflected electric field due to the spatial discontinuity.

The limiting value (E_{LB}) of {term3 + term4 + term5} in Equation 4.49 can be shown to be

$$E_{LB} = \frac{\omega_0}{\sqrt{\omega_0^2+\omega_p^2}}\left[\frac{\sqrt{\omega_0^2+\omega_p^2}-\omega_0}{\omega_p}\right]^2\cos\left(\sqrt{\omega_0^2+\omega_p^2}\,t\right). \tag{4.60}$$

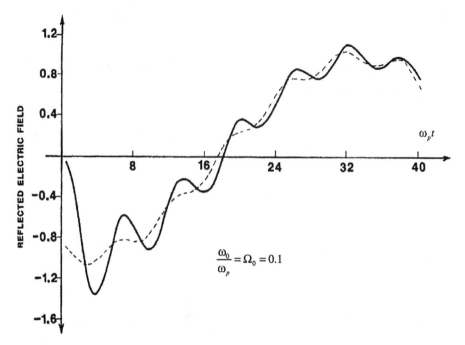

FIGURE 4.7   Reflected electric field vs. $\omega_p t$. The dotted curve shows the plot of the electric field E_L. $\Omega_0 = \omega_0/\omega_p = 0.1$. (From Kalluri, D. K., *IEEE Trans. Plasma Sci.*, 16, 11–16, 1988. With permission.)

This term E_{LB}, in the limiting value, is of a different frequency from the incident wave frequency and is due to the temporal discontinuity in the properties of the medium.[6] The multiplier of the cosine term on the right side of Equation 4.60 is the reflection coefficient R_B of Equation 4.8.

Figure 4.7 shows $\omega_p t$ vs. $E_{yr}(0,t)/E_0$ for $\Omega_0 = \omega_0/\omega_p = 0.1$. The dotted curve is the limiting value $E_L = E_{LA} + E_{LB}$ of this field Equation 4.57 + Equation 4.60, and is shown to indicate the period for which the transient aspect is important. For this curve, the incident wave frequency is less than the plasma frequency. Figure 4.8 shows $\omega_0 t$ vs. $E_{yr}(0,t)/E_0$ for $\Omega_0 = \omega_0/\omega_p = 2.5$. Here, the incident wave frequency is larger than the plasma frequency.

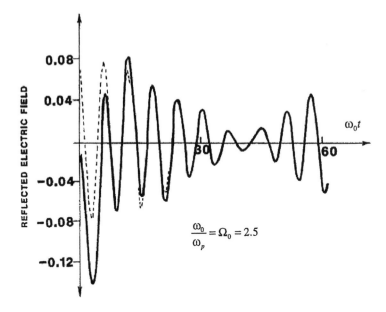

FIGURE 4.8.   Reflected electric field vs. $\omega_0 t$. The dotted curve shows E_L. $\Omega_0 = \omega_0/\omega_p = 2.5$. (From Kalluri, D. K., *IEEE Trans. Plasma Sci.*, 16, 11–16, 1988. With permission.)

REFERENCES

1. Felsen, B. L. and Whitman, G. M., Wave propagation in time-varying media, *IEEE Trans. Antennas Propag.*, AP-18, 242, 1970.
2. Fante, R. L., Transmission of electromagnetic waves into time-varying media, *IEEE Trans. Antennas Propag.*, AP-19, 417, 1971.
3. Jiang, C. L., Wave propagation and dipole radiation in a suddenly created plasma, *IEEE Trans. Antennas Propag.*, AP-23, 83, 1975.
4. Kalluri, D. K. and Prasad, R. C., Reflection of an electromagnetic wave from a suddenly created plasma half-space, in *Conf. Rec. Abstr., IEEE Int. Conf. Plasma Science*, San Diego, CA, 1983, 46.
5. Heald, M. A. and Wharton, C. B., *Plasma Diagnostics with Microwaves*, Wiley, New York, 1965.
6. Auld, B. A., Collins, J. H., and Zapp, H. R., Signal processing in a nonperiodically time-varying magnetoelastic medium, *Proc. IEEE*, 56, 258, 1968.
7. Kalluri, D. K., Numerical Laplace inversion of $(s^2 + a^2)^{-(n+1/2)} \ (s^2 + a^2)^{-m}$, *Int. J. Comput. Math.*, 19, 327, 1986.
8. Roberts, R. E. and Kaufman, H., *Table of Laplace Transforms*, W.B. Saunders, Philadelphia, 1966.

5 Switched Plasma Slab: B Wave Pulses*

5.1 INTRODUCTION

The previous chapter considered the switching of a plasma half-space and established that the reflected wave will have an A wave and a B wave. If the plasma is assumed to be lossless, the B wave at the upshifted frequency is present at $t = \infty$. This chapter examines the case of sudden creation of a lossless finite-extent plasma slab. The transient solution is obtained through the use of Laplace transforms. The solution is broken into two components: an A component, which in steady state has the same frequency as the incident wave frequency ω_0 and a B component, which has an upshifted frequency:

$$\omega_1 = \sqrt{\omega_0^2 + \omega_p^2} . \tag{5.1}$$

Here, ω_p is the plasma frequency of the switched plasma slab. Numerical results are presented to show the effect of the slab width on the reflected and transmitted waves.

For a finite slab width d, the B component decays and ultimately dies. This will be the case even if it is assumed that the created plasma slab is lossless. The conversion into B waves occurs at $t = 0$. After the medium change, there is no more wave conversion, and, therefore, there is no continuous energy flow from the original wave to the new plasma modes. In a lossless plasma, the B wave lasts longer and longer as d is increased. However, if the plasma is lossy, the amplitude of the B wave that emerges into free space after traveling in the lossy plasma will decrease as the slab width d increases.

A proper choice of ω_0, ω_p, and the slab width will yield a transmitted electromagnetic pulse of significant strength at the upshifted frequency.

5.2 DEVELOPMENT OF THE PROBLEM

The geometry of the problem is shown in Figure 5.1. Initially, for time $t < 0$, the entire space is considered to be free space. A uniform plane electromagnetic wave having a frequency of ω_0 rad/s and propagating in the positive z-direction is established over the entire space. This wave is called the incident wave. Its field components can be expressed in complex notation as

$$\mathbf{E}_i(z,t) = \hat{x} E_0 e^{j(\omega_0 t - k_0 z)}, \qquad t < 0, \tag{5.2}$$

* © 1992 American Institute of Physics. Reprinted with permission from Kalluri, D. K. and Goteti, V. R., Journal of Applied Physics, 72, 4575–4580, 1992. Chapter 5 is an adaptation of the reprint.

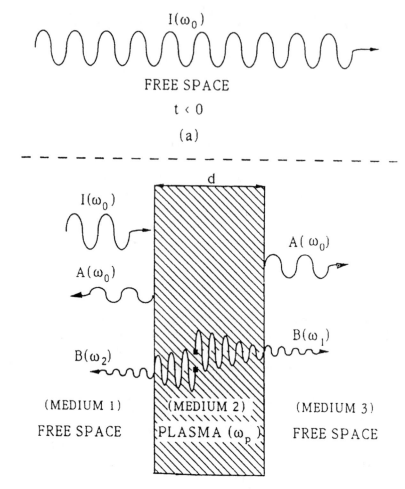

FIGURE 5.1 Effect of switching an isotropic plasma slab. Reflected and transmitted waves are sketched. A waves have the same frequency as the incident wave frequency (ω_0), but B waves have a new frequency $\omega_1 = (\omega_0^2 + \omega_p^2)^{1/2} = |\omega_2|$. (From Kalluri, D. K. and Goteti, V. R., *J. Appl. Phys.*, 72, 4575–4580, 1992. With permission.)

$$\mathbf{H}_i(z,t) = \hat{y} H_0 e^{j(\omega_0 t - k_0 z)}, \qquad t < 0. \tag{5.3}$$

Here, k_0 is the free-space wave number and $H_0 = E_0/\eta_0$, η_0 being the intrinsic impedance of free space. At time $t = 0$, an isotropic plasma medium of a constant plasma frequency ω_p is suddenly created between the planes $z = 0$ and $z = d$.

The problem features a temporal discontinuity resulting from a sudden temporal change in the properties of the medium and spatial discontinuities owing to the confinement of the plasma in space. Several interesting phenomena take place because of these discontinuities. The temporal discontinuity results in the creation of two new waves in the plasma medium, termed the B waves. They have the same wave number, k_0, as that of the incident wave but have a frequency that differs from that of the incident wave. On the other hand, spatial discontinuity results in the creation of two new waves in the plasma medium that have the same frequency, ω_0, as the incident wave but have a different wave number. These waves are termed the A waves. Out of each set, one of the two waves propagates in the positive z-direction whereas the other propagates in the negative z-direction. The composite effect of the creation of these waves is the production of reflected waves in free space for $z < 0$ and transmitted waves in free space for $z > d$ (Figure 5.1). The normalized (normalized with respect to E_0) electric field of the reflected wave due to the A wave is denoted by R_a at $z = 0$ and that of the transmitted wave is denoted by T_a at $z = d$. Similarly, the symbols R_b and T_b are used to denote the contributions from the B waves.

A solution to this problem is developed here using Laplace transforms with respect to the time variable t and ordinary differential equations with respect to the space variable z. The solution given here is restricted to the computation of R_a, R_b, T_a, and T_b.

5.3 TRANSIENT SOLUTION

For time $t > 0$, the space is divided into three zones, namely, free space up to $z = 0$, isotropic plasma from $z = 0$ to $z = d$, and, again, free space for $z > d$. The scalar electric fields in the three media are denoted by $E_1(z,t)$, $E_2(z,t)$, and $E_3(z,t)$, respectively. The partial differential equations for these fields are

$$\frac{\partial^2 E_1(z,t)}{\partial z^2} - \frac{1}{c^2}\frac{\partial^2 E_1(z,t)}{\partial t^2} = 0, \quad z < 0, \quad t > 0, \tag{5.4}$$

$$\frac{\partial^2 E_2(z,t)}{\partial z^2} - \frac{1}{c^2}\frac{\partial^2 E_2(z,t)}{\partial t^2} - \frac{\omega_p^2}{c^2}E_2(z,t) = 0, \quad 0 < z < d, t > 0, \tag{5.5}$$

$$\frac{\partial^2 E_3(z,t)}{\partial z^2} - \frac{1}{c^2}\frac{\partial^2 E_3(z,t)}{\partial t^2} = 0, \quad z > d, t > 0. \tag{5.6}$$

The newly created electrons are assumed to be at rest initially and are set in motion only for $t > 0$. This assumption is supported by the works of Jiang,[1] Kalluri,[2,3] and Goteti and Kalluri[4] and leads to the continuity of the electric and the magnetic field components over the temporal discontinuity at $t = 0$. The initial conditions are thus

$$E_1(t = 0^+) = E_2(t = 0^+) = E_3(t = 0^+) = E_i(t = 0^-) = E_0 \exp(-jk_0 z), \tag{5.7}$$

$$\frac{\partial E_1}{\partial t}(t = 0^+) = \frac{\partial E_2}{\partial t}(t = 0^+) = \frac{\partial E_3}{\partial t}(t = 0^+) = \frac{\partial E_i}{\partial t}(t = 0^-) = j\omega_0 E_0 \exp(-jk_0 z). \tag{5.8}$$

Using these initial conditions and taking the Laplace transforms of Equations 5.4 through 5.6 gives

$$\frac{d^2 E_1(z,s)}{dz^2} - q_1^2 E_1(z,s) = -\frac{(s+j\omega_0)}{c^2} E_0 \exp(-jk_0 z),$$ (5.9)

$$\frac{d^2 E_2(z,s)}{dz^2} - q_2^2 E_2(z,s) = -\frac{(s+j\omega_0)}{c^2} E_0 \exp(-jk_0 z),$$ (5.10)

$$\frac{d^2 E_3(z,s)}{dz^2} - q_1^2 E_3(z,s) = -\frac{(s+j\omega_0)}{c^2} E_0 \exp(-jk_0 z),$$ (5.11)

where $E_1(z,s)$, $E_2(z,s)$, and $E_3(z,s)$ are the Laplace transforms of $E_1(z,t)$, $E_2(z,t)$, and $E_3(z,t)$, respectively. Further,

$$q_1 = \frac{s}{c},$$ (5.12)

$$q_2 = \frac{\sqrt{(s^2 + \omega_p^2)}}{c}.$$ (5.13)

The solution to the ordinary differential equations given in Equations 5.4 through 5.6 can be written as

$$E_1(z,s) = A_1 \exp(q_1 z) + \frac{(s+j\omega_0)}{(s^2+\omega_0^2)} E_0 \exp(-jk_0 z), \quad z < 0,$$ (5.14)

$$E_2(z,s) = A_2 \exp(q_1 z) + A_3 \exp(-q_2 z) + \frac{(s+j\omega_0)}{(s^2+\omega_0^2+\omega_p^2)} E_0 \exp(-jk_0 z), \quad 0 < z < d, (5.15)$$

$$E_3(z,s) = A_4 \exp(-q_1 z) + \frac{(s+j\omega_0)}{(s^2+\omega_0^2)} E_0 \exp(-jk_0 z), \quad z > d.$$ (5.16)

The quantities A_1 through A_4 in Equations 5.14 through 5.16 are constants to be determined from the boundary conditions. The tangential components of the electric field and the magnetic field must be continuous at the $z = 0$ and $z = d$ interfaces. Thus,

$$E_1(0,s) = E_2(0,s),$$ (5.17)

$$\frac{dE_1(0,s)}{dz} = \frac{dE_2(0,s)}{dz}, \qquad (5.18)$$

$$E_2(d,s) = E_3(d,s), \qquad (5.19)$$

$$\frac{dE_2(d,s)}{dz} = \frac{dE_3(d,s)}{dz}. \qquad (5.20)$$

Substitution of the values of the constants in Equations 5.14 through 5.16 gives the complete description of the Laplace transforms of the fields in complex exponential form. When the incident wave is a harmonic wave of the form $E_0 \cos(\omega_0 t - k_0 z)$, it becomes necessary to take the real part on the right-hand side of Equations 5.14 through 5.16 to get the correct expressions for the fields. Following this procedure, the fields in the free-space zones, $z < 0$ and $z > d$ are obtained and given below:

$$E_1(z,s) = A_{1R}(s)\exp(q_1 z) + E_0\frac{\left[s\cos(k_0 z) + \omega_0 \sin(k_0 z)\right]}{\left(s^2 + \omega_0^2\right)}, \quad z > 0, \qquad (5.21)$$

$$E_3(z,s) = A_{4R}(s)\exp\left[-q_1(z-d)\right] + E_0\frac{\left[s\cos(k_0 z) + \omega_0 \sin(k_0 z)\right]}{\left(s^2 + \omega_0^2\right)}, \quad z > d, \qquad (5.22)$$

where A_{1R} and A_{4R} are the real parts of A_1 and A_4, respectively.

The second term on the right-hand side of Equation 5.21 is the Laplace transform of the incident wave. Hence, the first term corresponds to the field of the reflected waves in the free-space zone $z < 0$. Specifically, the inverse Laplace transform of $A_{1R}(s)$ gives the time variation of the fields of the $z < 0$, and the fields at any distance $z < 0$ can be obtained by replacing t with $(t + z/c)$. Similarly, the inverse Laplace transform of $A_{4R}(s)$ gives the fields of the transmitted waves at $z = d$, and replacement of t with $(t - (z - d)/c)$ gives the fields for any $z > d$. The complete expressions for $A_{1R}(s)$ and $A_{4R}(s)$ are given below:

$$\frac{A_{1R}}{E_0}(s) = R_A(s) + R_B(s), \qquad (5.23)$$

$$R_A(s) = \frac{\omega_p^2\left[\exp(q_2 d) - \exp(-q_2 d)\right]}{DR}\frac{s}{\left(s^2 + \omega_0^2\right)}, \qquad (5.24)$$

$$R_B(s) = \frac{\omega_p^2 \left[\exp(q_2 d) + \exp(-q_2 d) - 2\cos(k_0 d) \right]}{DR} \frac{\sqrt{s^2 + \omega_p^2}}{\left(s^2 + \omega_0^2 + \omega_p^2 \right)}, \tag{5.25}$$

$$\frac{A_{4R}}{E_0}(s) = T_A(s) + T_B(s) - E_0 \frac{s\cos(k_0 d) + \omega_0 \sin(k_0 d)}{s^2 + \omega_0^2}, \tag{5.26}$$

$$T_A(s) = -\frac{4s^2 \sqrt{s^2 + \omega_p^2}}{DR\left(s^2 + \omega_0^2 \right)}, \tag{5.27}$$

$$T_B(s) = \frac{\omega_0 \sin(k_0 d)}{\left(s^2 + \omega_0^2 + \omega_p^2 \right)} + \frac{\sqrt{s^2 + \omega_p^2}}{\left(s^2 + \omega_0^2 + \omega_p^2 \right)} \left[T_{B1}(s) - T_{B2}(s) \right], \tag{5.28}$$

$$T_{B1}(s) = \frac{2\left(2s^2 + \omega_p^2 \right)}{DR}, \tag{5.29}$$

$$T_{B2}(s) = \cos(k_0 d) \left(\frac{\left[s - \sqrt{\left(s^2 + \omega_p^2 \right)} \right]^2 \exp(-q_2 d)}{DR} \right)$$

$$+ \cos(k_0 d) \left(\frac{\left[s + \sqrt{\left(s^2 + \omega_p^2 \right)} \right]^2 \exp(q_2 d)}{DR} \right), \tag{5.30}$$

where

$$DR = \left(s - \sqrt{s^2 + \omega_p^2} \right)^2 \exp(-q_2 d) - \left(s + \sqrt{s^2 + \omega_p^2} \right)^2 \exp(q_2 d) \tag{5.31}$$

and q_1 and q_2 are given by Equations 5.12 and 5.13, respectively.

An observation of Equation 5.24 shows that $R_A(s)$ corresponds to the field of the reflected wave that has a frequency ω_0 in steady state. This component arises as a result of reflection of the incident wave from the space boundary at $z = 0$. The term $R_B(s)$ refers to the transient effects involved in the partial transmission of the negatively propagating B wave across the $z = 0$ space boundary into the free-space zone $z < 0$. Similarly, in Equations 5.26, $T_A(s)$ refers to the electric field of the transmitted wave at the $z = d$ space boundary. Its steady-state frequency is ω_0. $T_A(s)$ is due to the spatial discontinuity in the properties of the medium at $z = d$. $T_B(s)$ refers to the partial transmission of the positively propagating B wave across the $z = d$ space boundary into the free-space zone $z > d$. Thus, based on the observation of the location of the complex poles in Equations 5.24 and 5.27, it can be concluded that quantities with the suffix A refer to the effects of spatial discontinuities and those with the suffix B refer to the effects of temporal discontinuity in the properties of the medium. This identification is further confirmed from an analysis of the steady-state solution given in Section 5.5.

5.4 DEGENERATE CASE

The case of imposition of a semi-infinite plasma (d tending to infinity) is examined here. When the slab width d tends to infinity, the fields of the reflected waves given in Equations 5.24 and 5.25 can be shown to reduce to the following:

$$R_A(s) = -\frac{\omega_p^2}{\left(s + \sqrt{s^2 + \omega_p^2}\right)^2}\frac{s}{\left(s^2 + \omega_0^2\right)}, \tag{5.32}$$

$$R_B(s) = -\frac{\omega_p^2}{\left(s + \sqrt{s^2 + \omega_p^2}\right)^2}\frac{\sqrt{s^2 + \omega_p^2}}{\left(s^2 + \omega_0^2 + \omega_p^2\right)}. \tag{5.33}$$

These expressions agree with the results obtained by Kalluri[2] in his analysis of wave propagation due to a suddenly imposed semi-infinite plasma medium.

5.5 *A* COMPONENT FROM STEADY-STATE SOLUTION

Based on the assumption of the following form of waves in the three media, this steady-state boundary value problem is solved:

Medium 1 ($z < 0$): Free Space

Incident wave:

$$E_1(z,t) = E_0 \exp j(\omega_0 t - k_0 z), \tag{5.34}$$

Reflected wave:

$$E_R(z,t) = B_1 \exp j(\omega_0 t + k_0 z). \tag{5.35}$$

Medium 2 (0 < z < d): Isotropic Plasma

Transmitted wave:

$$E_p^+(z,t) = B_2 \exp\left[j(\omega_0 t - \beta z) \right], \tag{5.36}$$

Reflected wave:

$$E_p^-(z,t) = B_3 \exp\left[j(\omega_0 t + \beta z) \right]. \tag{5.37}$$

Medium 3 (z > d): Free Space

Transmitted wave:

$$E_T(z,t) = B_4 \exp\left[j(\omega_0 t - k_0 z) \right], \tag{5.38}$$

where β is the propagation constant in the plasma medium and is given by

$$\beta = \frac{\sqrt{\left(\omega_0^2 - \omega_p^2\right)}}{c}. \tag{5.39}$$

By using the boundary conditions at $z = 0$ and at $z = d$, the following expressions for the B_1 and B_4 can be obtained:

$$\frac{B_1}{E_0} = \frac{\left(1 - a^2\right)\left[\exp(j\beta d) - \exp(-j\beta d)\right]}{(1+a)^2 \exp(j\beta d) - (1-a)^2 \exp(-j\beta d)}, \tag{5.40}$$

$$\frac{B_4}{E_0} = \frac{4a \exp(jk_0 d)}{(1+a)^2 \exp(j\beta d) - (1-a)^2 \exp(-j\beta d)}, \tag{5.41}$$

where

$$a = \frac{\sqrt{\left(\omega_0^2 - \omega_p^2\right)}}{\omega_0}. \tag{5.42}$$

B_1 and B_4 obtained in Equations 5.40 and 5.41 are the frequency response coefficients of the fields. From this steady-state solution, it is possible to obtain the Laplace transforms of the transient fields of the reflected wave and the transmitted wave in free space when the incident wave is of the form $E_0 \cos(\omega_0 t - k_0 z)$. The steps are

1. Replace $j\omega_0$ in Equations 5.40 and 5.41 with s;
2. Multiply the resulting expression with $s/(s^2 + \omega_0^2)$, the Laplace transform of $\cos(\omega_0 t)$;
3. Multiply B_1 with the operator $\exp(q_1 z)$ and B_4 with the operator $\exp(-q_1 z)$ to account for the propagation of the reflected and transmitted waves in free space.

This procedure gives the following expressions:

$$\frac{E_R}{E_0}(z,s) = \frac{\omega_p^2 \left[\exp(q_2 d) - \exp(-q_2 d)\right]}{DR} \frac{s}{\left(s + \omega_0^2\right)} \exp(q_2 z), \qquad (5.43)$$

$$\frac{E_T}{E_0}(z,s) = \frac{4s^2 \sqrt{\left(s^2 + \omega_p^2\right)}}{DR\left(s^2 + \omega_0^2\right)} \exp\left[-q_1(z - d)\right], \qquad (5.44)$$

where DR, q_1, and q_2 are given in Equation 5.31. Comparison of Equation 5.43 with 5.24 and Equation 5.44 with 5.27 shows that $R_A(s)$ and $T_A(s)$ are indeed the components resulting from the effects of spatial discontinuities in the properties of the medium. Thus, the split-up given in Equations 5.23 and 5.26 into components arising out of spatial discontinuities and temporal discontinuity is correct and complete. Hence, the inverse Laplace transforms of $R_A(s)$ and $T_A(s)$ can be expected to merge into the steady-state solution given in Equations 5.40 and 5.47. Proceeding on the same lines, the steady-state behavior of $R_B(s)$ and $T_B(s)$ can be obtained. This requires assuming two waves in the plasma medium having the same frequency ω_1, but propagating in opposite directions, and analyzing the reflection and transmission aspects into free space across the $z = 0$ and $z = d$ space boundaries. The amplitudes of these waves in free space can be shown to be zero in steady state for finite width of the plasma slab. Thus, the steady-state values of $R_b(t)$ and $T_b(t)$ can be expected to be zero for finite values of d.

Numerical Laplace inversion[5] of the expressions given in Equations 5.24, 5.25, 5.27, and 5.28 is performed, and these observations have been confirmed in Section 5.6.

5.6   NUMERICAL RESULTS

The numerical results are presented in normalized form. The frequency variables are normalized by taking $\omega_p = 1$, and d is normalized with respect to λ_p (free-space wavelength corresponding to plasma frequency). The electric fields of the reflected and transmitted waves are normalized with respect to the strength of the electric field (E_0) of the incident wave.

Figure 5.2 presents reflection coefficients at $z = 0$ vs. t for the semi-infinite problem. The results are obtained by performing numerical Laplace inversion of Equations 5.32 and 5.33. $R_a(t)$ shows a frequency of ω_0 and $R_b(t)$ a frequency of

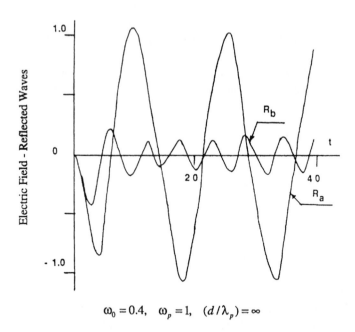

$$\omega_0 = 0.4, \quad \omega_p = 1, \quad (d/\lambda_p) = \infty$$

FIGURE 5.2   Electric field of reflected waves vs. t for the case of switching a lossless plasma half-space. The results are presented in normalized form. The frequency variables are normalized by taking $\omega_p = 1$. $R_a(t)$ and $R_b(t)$ are the electric fields at $z = 0$ normalized with respect to the strength of the electric field (E_0) of the incident wave. $R_b(t)$ has a steady-state frequency $\omega_1 = (\omega_0^2 + \omega_p^2)^{1/2}$ and a nonzero steady-state amplitude. (From Kalluri, D. K. and Goteti, V. R., *J. Appl. Phys.*, 72, 4575–4580, 1992. With permission.)

$\omega_1 = \sqrt{\omega_0^2 + \omega_p^2}$ as t approaches infinity. The parameter ω_0 is chosen as 0.4 since it is known from previous work[2] that R_b will have maximum amplitude at about this value of ω_0. In this idealized model of lossless plasma R_b will persist forever, and its amplitude has a nonzero steady-state value. This phenomenon may be physically explained in the following way;[2] one of the B waves generated at $t = 0$ and $z = \infty$ propagates in the negative z-direction and emerges into medium 1 at $t = \infty$.

For a finite value of d, the B component should decay since the conversion into B waves occurs at $t = 0$. After the medium change, there is no more wave conversion; therefore, there is no continuous energy flow from the original wave to the new plasma modes. Figure 5.3 illustrates this point for the parameter $d/\lambda_p = 0.5$. The B wave that is generated at $t = 0$ and is traveling along positive z-direction emerges into medium 3 and gives rise to T_b. Since $\omega_0 < \omega_p$, the incident wave frequency is in the stop band of the ordinary wave. T_a is due to tunneling (see Section 2.5) of the incident wave through the plasma slab and will become significantly weaker as the slab width increases. Medium 3 then will not have signals at the frequency of the incident wave. The above point is illustrated by progressively increasing d/λ_p. The results for the transmitted waves are shown in Figures 5.4, 5.5, and 5.6. The parameter ω_0 is chosen as 0.8. This frequency is still in the stop band, but results in a

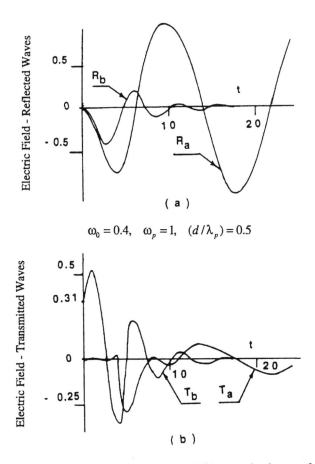

$$\omega_0 = 0.4, \quad \omega_p = 1, \quad (d/\lambda_p) = 0.5$$

FIGURE 5.3 Electric field of (a) reflected waves and (b) transmitted waves for the case of switching a lossless plasma slab of width d. The width is normalized with respect to λ_p (free-space wavelength corresponding to plasma frequency). $T_a(t)$ and $T_b(t)$ are normalized electric field variables at $z = d$. The B waves decay even if the plasma slab is assumed to be lossless. (From Kalluri, D. K. and Goteti, V. R., *J. Appl. Phys.*, 72, 4575–4580, 1992. With permission.)

stronger B component in the transmitted wave. Figure 5.6 shows only a B component in the transmitted wave. The A component has negligible amplitude for $d/\lambda_p = 2.0$. Thus, it is possible to obtain an upshifted signal[6] and suppress the original signal[6] by rapid creation of a plasma slab.

Electromagnetics of Complex Media

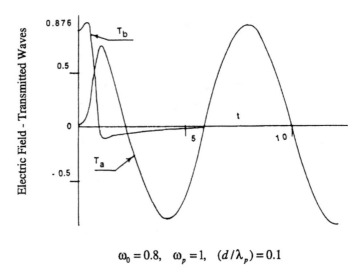

$$\omega_0 = 0.8, \quad \omega_p = 1, \quad (d/\lambda_p) = 0.1$$

FIGURE 5.4 $T_a(t)$ and $T_b(t)$ vs. t for $d/\lambda_p = 0.1$. Since $\omega_0 < \omega_p$, T_a is due to tunneling of the incident wave through the plasma slab. It will become significantly weaker as the slab width increases. (From Kalluri, D. K. and Goteti, V. R., *J. Appl. Phys.*, 72, 4575–4580, 1992. With permission.)

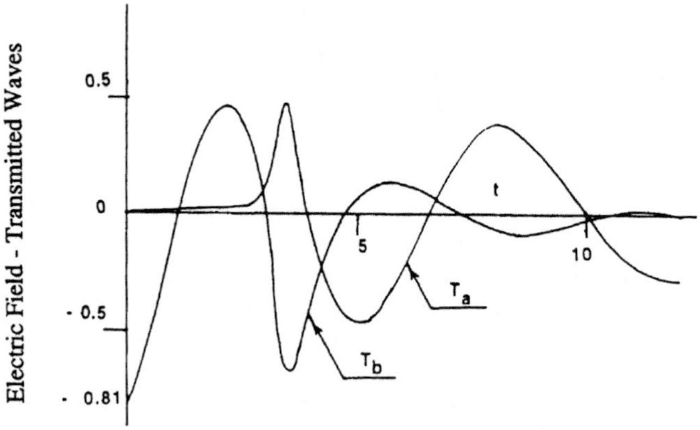

$$\omega_0 = 0.8, \quad \omega_p = 1, \quad (d/\lambda_p) = 0.5$$

FIGURE 5.5 $T_a(t)$ and $T_b(t)$ vs. t for $d/\lambda_p = 0.5$. (From Kalluri, D. K. and Goteti, V. R., *J. Appl. Phys.*, 72, 4575–4580, 1992. With permission.)

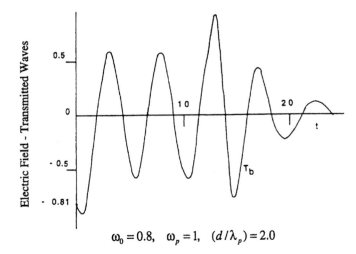

$$\omega_0 = 0.8, \quad \omega_p = 1, \quad (d/\lambda_p) = 2.0$$

FIGURE 5.6 $T_b(t)$ vs. t for d/λ_p. $T_a(t)$ is negligible. (From Kalluri, D. K. and Goteti, V. R., *J. Appl. Phys.*, 72, 4575–4580, 1992. With permission.)

REFERENCES

1. Jiang, C. L., Wave propagation and dipole radiation in a suddenly created plasma, *IEEE Trans. Antennas Propag*, AP-23, 83, 1975.
2. Kalluri, D. K., On reflection from a suddenly created plasma half-space: transient solution, *IEEE Trans. Plasma Sci.*, 16, 11, 1988.
3. Kalluri, D. K., Effect of switching a magnetoplasma medium on a traveling wave: longitudinal propagation, *IEEE Trans. Antennas Propag.*, AP-37, 1638, 1989.
4. Goteti, V. R. and Kalluri, D. K., Wave propagation in a switched magnetoplasma medium: transverse propagation, *Radio Sci.*, 25, 61, 1990.
5. *IMSL Library Reference Manual*, Vol. 2, IMSL, Inc., Houston, 1982, FLINV-1.
6. Wilks, S. C., Dawson, J. M., and Mori, W. B., Frequency up-conversion of electromagnetic radiation with use of an overdense plasma, *Phys. Rev. Lett.*, 61, 337, 1988.

6 Magnetoplasma Medium: L, R, O, and X Waves

6.1 INTRODUCTION

Plasma medium in the presence of a static magnetic field behaves like an anisotropic dielectric. Therefore, the rules of electromagnetic wave propagation are the same as those of the light waves in crystals. Of course, in addition, account has to be taken of the dispersive nature of the plasma medium.

The well-established *magnetoionic theory*[1-3] is concerned with the study of wave propagation of an arbitrarily polarized plane wave in a cold anisotropic plasma, where the direction of phase propagation of the wave is at an arbitrary angle to the direction of the static magnetic field. As the wave travels in such a medium, the polarization state continuously changes. However, there are specific normal modes of propagation in which the state of polarization is unaltered. Waves with the left (*L* wave) or the right (*R* wave) circular polarization are the normal modes in the case of phase propagation along the static magnetic field. Such propagation is labeled *longitudinal propagation*. The ordinary wave (*O* wave) and the extraordinary wave (*X* wave) are the normal modes for *transverse propagation*. The properties of these waves are explored in the Sections 6.3 through 6.5.

6.2 BASIC FIELD EQUATIONS FOR A COLD ANISOTROPIC PLASMA MEDIUM

In the presence of a static magnetic field $\mathbf{B}_0$, the force equation (Equation 1.3) needs modification because of the additional magnetic force term:

$$m\frac{d\mathbf{v}}{dt} = -q\left[\mathbf{E} + \mathbf{v} \times \mathbf{B}_0\right]. \tag{6.1}$$

The corresponding modification of Equation 1.10 is given by

$$\frac{d\mathbf{J}}{dt} = \varepsilon_0 \omega_p^2(\mathbf{r},t)\mathbf{E} - \mathbf{J} \times \omega_b, \tag{6.2}$$

where

$$\omega_b = \frac{q\mathbf{B}_0}{m} = \omega_b \hat{\mathbf{B}}_0. \tag{6.3}$$

In the above, $\hat{\mathbf{B}}_0$ is a unit vector in the direction of $\mathbf{B}_0$ and ω_b is the absolute value of the electron gyrofrequency.

Equations 1.1 and 1.2 repeated here for convenience,

$$\nabla \times \mathbf{E} = -\mu_0 \frac{\partial \mathbf{H}}{\partial t},$$

(6.4)

and

$$\nabla \times \mathbf{H} = \varepsilon_0 \frac{\partial \mathbf{E}}{\partial t} + \mathbf{J},$$

(6.5)

and Equation 6.2 are the basic equations that will be used in discussing electromagnetic wave transformation (EMW) by a magnetized cold plasma.

By taking the curl of Equation 1.1 and eliminating $\mathbf{H}$, the wave equation for $\mathbf{E}$ can be derived:

$$\nabla^2 \mathbf{E} - \nabla(\nabla \cdot \mathbf{E}) - \frac{1}{c^2} \frac{\partial^2 \mathbf{E}}{\partial t^2} - \frac{1}{c^2} \omega_p^2(\mathbf{r},t)\mathbf{E} + \mu_0 \mathbf{J} \times \boldsymbol{\omega}_b = 0.$$

(6.6)

Similar efforts will lead to a wave equation for the magnetic field:

$$\nabla^2 \dot{\mathbf{H}} - \frac{1}{c^2} \frac{\partial^2 \dot{\mathbf{H}}}{\partial t^2} - \frac{1}{c^2} \omega_p^2(\mathbf{r},t)\dot{\mathbf{H}} + \varepsilon_0 \nabla \omega_p^2(\mathbf{r},t) \times \mathbf{E} + \nabla \times (\mathbf{J} \times \mathbf{E}) = 0,$$

(6.7)

where

$$\dot{\mathbf{H}} = \frac{\partial \mathbf{H}}{\partial t}.$$

(6.8)

If ω_p^2 and ω_b vary only with t, Equation 6.7 becomes

$$\nabla^2 \dot{\mathbf{H}} - \frac{1}{c^2} \frac{\partial^2 \dot{\mathbf{H}}}{\partial t^2} - \frac{1}{c^2} \omega_p^2(t)\dot{\mathbf{H}} + \varepsilon_0 \omega_b(t)\left(\nabla \cdot \dot{\mathbf{E}}\right) = 0.$$

(6.9)

Equations 6.6 and 6.7 involve more than one field variable. It is possible to convert them into higher-order equations in one variable. In any case, it is difficult to obtain meaningful analytical solutions to these higher-order vector partial differential equations. The equations in this section are useful in developing numerical methods.

Considered next, as in the isotropic case, are simple solutions to particular cases where one parameter or one aspect is highlighted at a time. These solutions will serve as building blocks for the more-involved problems.

6.3   ONE-DIMENSIONAL EQUATIONS: LONGITUDINAL PROPAGATION, *L* AND *R* WAVES

The particular case is considered where (1) the variables are functions of one spatial coordinate only, say, the z-coordinate, (2) the electric field is circularly polarized, (3) the static magnetic field is z-directed, and (4) the variables are denoted by

$$\mathbf{E} = (\hat{x} \mp j\hat{y})E(z,t), \tag{6.10}$$

$$\mathbf{H} = (\pm j\hat{x} + \hat{y})H(z,t), \tag{6.11}$$

$$\mathbf{J} = (\hat{x} \mp j\hat{y})J(z,t), \tag{6.12}$$

$$\omega_p^2 = \omega_p^2(z,t), \tag{6.13}$$

$$\boldsymbol{\omega}_b = \hat{z}\omega_b(z,t). \tag{6.14}$$

The basic equations for E, H, and J take the following simple form:

$$\frac{\partial E}{\partial z} = -\mu_0 \frac{\partial H}{\partial t}, \tag{6.15}$$

$$-\frac{\partial H}{\partial z} = \varepsilon_0 \frac{\partial E}{\partial t} + J, \tag{6.16}$$

$$\frac{dJ}{dt} = \varepsilon_0 \omega_p^2(z,t)E \pm j\omega_b(z,t)J, \tag{6.17}$$

$$\frac{\partial^2 E}{\partial z^2} - \frac{1}{c^2}\frac{\partial^2 E}{\partial t^2} - \frac{1}{c^2}\omega_p^2(z,t)E \mp \frac{j}{c^2}\omega_b(z,t)\frac{\partial E}{\partial t} \mp j\,\mu_0\omega_b(z,t)\frac{\partial H}{\partial z} = 0, \tag{6.18}$$

$$\frac{\partial^2 H}{\partial z^2} - \frac{1}{c^2}\frac{\partial^2 H}{\partial t^2} - \frac{1}{c^2}\omega_p^2(z,t)\dot{H} + \varepsilon_0 \frac{\partial \omega_b(z,t)}{\partial z}E \mp j\omega_b(z,t)\frac{\partial^2 H}{\partial z^2}$$

$$\pm \frac{1}{c^2}j\omega_b(z,t)\frac{\partial^2 H}{\partial t^2} \mp j\frac{\partial \omega_b(z,t)}{\partial z}\frac{\partial H}{\partial z} \mp j\varepsilon_0\frac{\partial \omega_b(z,t)}{\partial z}\frac{\partial E}{\partial t} = 0. \tag{6.19}$$

If ω_p and ω_b are functions of time only, Equation 6.19 reduces to

$$\frac{\partial^2 \dot{H}}{\partial z^2} - \frac{1}{c^2}\frac{\partial^2 \dot{H}}{\partial t^2} - \frac{1}{c^2}\omega_p^2(t)\dot{H} \mp j\omega_b(t)\frac{\partial^2 H}{\partial z^2} \pm \frac{1}{c^2}j\omega_b(t)\frac{\partial^2 H}{\partial t^2} = 0. \tag{6.20}$$

Next, plane wave solutions in a homogeneous, time-invariant unbounded magneto-plasma medium are sought, i.e.,

$$f(z,t) = \exp[j(\omega t - kz)],$$

(6.21)

$$\omega_p^2(z,t) = \omega_p^2,$$

(6.22)

$$\omega_b(z,t) = \omega_b,$$

(6.23)

where f stands for any of the field variables E, H, or J. From Equations 6.16 and 6.17, it is shown that

$$-jkH = j\omega \, \varepsilon_0 \varepsilon_{pR,L} E,$$

(6.24)

where

$$\varepsilon_{pR,L} = 1 - \frac{\omega_p^2}{\omega(\omega \mp \omega_b)}.$$

(6.25)

This shows clearly that the magnetized plasma, in this case, can be modeled as a dielectric with the dielectric constant given by Equation 6.25. The dispersion relation is obtained from

$$k^2 = \frac{\omega^2}{c^2} \varepsilon_{pR,L},$$

(6.26)

which when expanded gives

$$\omega^3 \mp \omega_b \omega^2 - \left(k^2 c^2 + \omega_p^2\right)\omega \pm k^2 c^2 \omega_b = 0.$$

(6.27)

The expression for the dielectric constant can be written in an alternative fashion in terms of ω_{c1} and ω_{c2}, which are called cutoff frequencies: the dispersion relation can also be recast in terms of the cutoff frequencies.

$$\varepsilon_{pR,L} = \frac{(\omega \pm \omega_{c1})(\omega \mp \omega_{c2})}{\omega(\omega \mp \omega_b)},$$

(6.28)

$$\omega_{c1,c2} = \mp\frac{\omega_b}{2} + \sqrt{\left(\frac{\omega_b}{2}\right)^2 + \omega_p^2},$$

(6.29)

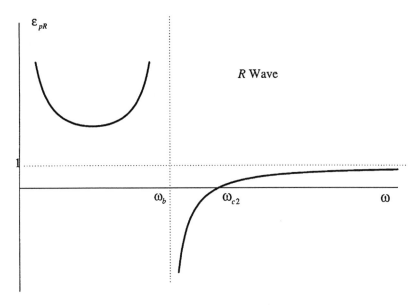

FIGURE 6.1 Dielectric constant for an R wave.

$$k_{R,L}^2 c^2 = \omega^2 \varepsilon_{pR,L} = \frac{\omega(\omega \pm \omega_{c1})(\omega \mp \omega_{c2})}{(\omega \mp \omega_b)}. \qquad (6.30)$$

In the above, the top sign is for the right circular polarization (R wave) and the bottom sign for the left circular polarization. Figure 6.1 gives ε_p vs. ω and Figure 6.2 the ω–k diagram, respectively, for the R wave. Figure 6.3 and Figure 6.4 do the same for the L wave. R and L waves propagate in the direction of the static magnetic field, without any change in the polarization state of the wave, and are called the characteristic waves of longitudinal propagation. For these waves, the medium behaves like an isotropic plasma except that the dielectric constant ε_p is influenced by the strength of the static magnetic field. Particular attention is drawn to Figure 6.1, which shows that the dielectric constant $\varepsilon_p > 1$ for the R wave in the frequency band $0 < \omega < \omega_b$. This mode of propagation is called the *whistler mode* in the literature on ionospheric physics and the helicon mode in the literature on solid-state plasma. Chapter 9 deals with the transformation of the whistler wave by a transient magnetoplasma and the consequences of such a transformation. Note that an isotropic plasma does not support a whistler wave.

The longitudinal propagation of a linearly polarized wave is accompanied by *Faraday rotation* of the plane of polarization. This phenomenon is easily explained in terms of the propagation of the R and L characteristic waves. A linearly polarized wave is the superposition of R and L waves each of which propagates without change of its polarization state but each with a different phase velocity. (Note from Equation 6.25 that ε_p for a given ω, ω_p, and ω_b is different for R and L waves because

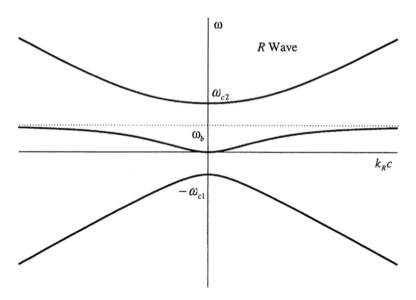

FIGURE 6.2 ω–k diagram for an R wave.

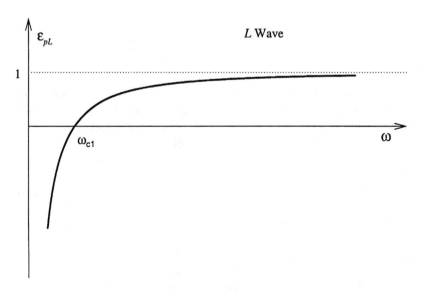

FIGURE 6.3   Dielectric constant for an L wave.

of the sign difference in the denominator). The result of combining the two waves after traveling a distance d is a linearly polarized wave again, but with the plane of polarization rotated by an angle Ψ, given by

$$\Psi = \tfrac{1}{2}\big(k_L - k_R\big)d. \tag{6.31}$$

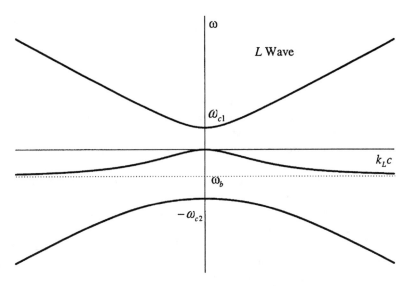

FIGURE 6.4 ω–k diagram for an L wave.

The high value of ε_p in the lower-frequency part of the whistler mode may be demonstrated by the following calculation given in Reference 2; for $f_p = 10^{14}$ Hz, $f_b = 10^{10}$ Hz and $f = 10$ Hz, $\varepsilon_p = 9 \times 10^{16}$. The wavelength of the signal in the plasma is only 10 cm. See References 4 through 6 for literature on the associated phenomena.

The resonance of the R wave at $\omega = \omega_b$ is of special significance. Around this frequency it can be shown that even a low-loss plasma strongly absorbs the energy of a source electromagnetic wave and heats the plasma. This effect is the basis of radio frequency heating of *fusion plasmas*.[7] It is also used to determine experimentally the effective mass of an electron in a crystal.[8]

For the L wave, Figure 6.3 does not show any resonance effect. In fact, the L wave has resonance at the ion gyrofrequency. In the Lorentz plasma model the ion motion is neglected. A simple modification, by including the ion equation of motion, will extend the EMW transformation theory to a low-frequency source wave.[2,9-11]

6.4 ONE-DIMENSIONAL EQUATIONS:
TRANSVERSE PROPAGATION, *O* WAVE

Next an electric field linearly polarized in the y-direction is considered, i.e.,

$$\mathbf{E} = \hat{y}E(z,t), \tag{6.32}$$

$$\mathbf{H} = -\hat{x}H(z,t), \tag{6.33}$$

$$\mathbf{J} = \hat{y}J(z,t), \tag{6.34}$$

$$\omega_p^2 = \omega_p^2(z,t),\tag{6.35}$$

$$\boldsymbol{\omega}_b = \hat{y}\omega_b(z,t).\tag{6.36}$$

The last term on the right side of Equation 6.2 is zero since the current density and the static magnetic field are in the same direction. The equations then are no different from those of the isotropic case, and the static magnetic field has no effect. The electrons move in the direction of the electric field and give rise to a current density in the plasma in the same direction as that of the static magnetic field. In such a case, the electrons do not experience any magnetic force, their orbit is not bent, and they continue to move in the direction of the electric field. The one-dimensional solution for a plane wave in such a medium is called an *ordinary wave* or *O* wave. Its characteristics are the same as those discussed in previous chapters. In this case, unlike in the case considered in Section 6.2, the direction of phase propagation is perpendicular to the direction of the static magnetic field, and, hence, this case comes under the label of transverse propagation.

6.5 ONE-DIMENSIONAL SOLUTION: TRANSVERSE PROPAGATION, *X* WAVE

The more difficult case of transverse propagation occurs when the electric field is normal to the static magnetic field. The trajectories of the electrons that start moving in the direction of the electric field get altered and bent as a result of the magnetic force. Such a motion gives rise to an additional component of the current density in the direction of phase propagation, and to obtain a self-consistent solution it must be assumed that a component of the electric field also exists in the direction of phase propagation. Let

$$\mathbf{E} = \hat{x}E_x(z,t) + \hat{z}E_z(z,t),\tag{6.37}$$

$$\mathbf{H} = \hat{y}H(z,t),\tag{6.38}$$

$$\mathbf{J} = \hat{x}J_x(z,t) + \hat{z}J_z(z,t),\tag{6.39}$$

$$\omega_p^2 = \omega_p^2(z,t),\tag{6.40}$$

$$\boldsymbol{\omega}_b = \hat{y}\omega_b(z,t).\tag{6.41}$$

The basic equations for *E*, *H*, and *J* take the following form:

$$\frac{\partial E_x}{\partial z} = -\mu_0 \frac{\partial H}{\partial t},\tag{6.42}$$

$$-\frac{\partial H}{\partial z} = \varepsilon_0 \frac{\partial E_x}{\partial t} + J_x, \tag{6.43}$$

$$\varepsilon_0 \frac{\partial E_z}{\partial t} = -J_z, \tag{6.44}$$

$$\frac{dJ_x}{dt} = \varepsilon_0^2 \omega_p^2(z,t)E_x + \omega_b(z,t)J_z, \tag{6.45}$$

$$\frac{dJ_z}{dt} = \varepsilon_0^2 \omega_p^2(z,t)E_z - \omega_b(z,t)J_x. \tag{6.46}$$

The plane wave solution in a homogeneous, time-invariant, unbounded magneto-plasma medium is again sought by applying Equation 6.21 to the set of Equations 6.42 through 6.46:

$$-jkE_x = -\mu_0 j\omega H, \tag{6.47}$$

$$-jkH = j\omega\varepsilon_0 E_x + J_x, \tag{6.48}$$

$$\varepsilon_0 j\omega E_z = -J_z, \tag{6.49}$$

$$j\omega J_x = \varepsilon_0 \omega_p^2 E_x + \omega_b J_z, \tag{6.50}$$

$$j\omega J_z = \varepsilon_0 \omega_p^2 E_z - \omega_b J_x. \tag{6.51}$$

From Equations 6.49 through 6.51, the following relation between J_x and E_x is obtained:

$$J_x = \varepsilon_0 \frac{\omega_p^2}{\left[1 - \dfrac{\omega_b^2}{\omega^2 - \omega_p^2}\right]} \frac{E_x}{j\omega}. \tag{6.52}$$

By substituting Equation 6.52 in Equation 6.48,

$$-jkH = j\omega\varepsilon_0\varepsilon_{pX}E_x, \tag{6.53}$$

$$\varepsilon_{pX} = 1 - \frac{\omega_p^2/\omega^2}{\left[1 - \dfrac{\omega_b^2}{\omega^2 - \omega_p^2}\right]}. \tag{6.54}$$

An alternative expression for ε_{pX} can be written in terms of ω_{c1}, ω_{c2}, and ω_{uh}:

$$\varepsilon_{pX} = \frac{\left(\omega^2 - \omega_{c1}^2\right)\left(\omega^2 - \omega_{c2}^2\right)}{\omega^2\left(\omega^2 - \omega_{uh}^2\right)}. \tag{6.55}$$

The cutoff frequencies ω_{c1} and ω_{c2} have been defined earlier, and ω_{uh} is the upper hybrid frequency:

$$\omega_{uh}^2 = \omega_p^2 + \omega_b^2. \tag{6.56}$$

The dispersion relation $k^2 = (\omega^2/c^2)\varepsilon_{pX}$, when expanded, gives

$$\omega^4 - \left(k^2c^2 + \omega_b^2 + 2\omega_p^2\right)\omega^2 + \left[\omega_p^4 + k^2c^2\left(\omega_b^2 + \omega_p^2\right)\right] = 0. \tag{6.57}$$

In obtaining Equation 6.57, the expression for ε_{pX} given by Equation 6.54 is used. For the purpose of sketching the ω–k diagram, it is more convenient to use Equation 6.54 for ε_{pX} and write the dispersion relation as

$$k^2c^2 = \frac{\left(\omega^2 - \omega_{c1}^2\right)\left(\omega^2 - \omega_{c2}^2\right)}{\left(\omega^2 - \omega_{uh}^2\right)}. \tag{6.58}$$

Figures 6.5 and 6.6 sketch ε_{pX} vs. ω and ω vs. kc, respectively. By substituting Equations 6.49 and 6.52 in Equation 6.51, the ratio of E_z to E_x can be found:

$$\frac{E_z}{E_x} = -j\omega \frac{\omega_b}{\omega^2 - \omega_p^2}\left(\varepsilon_{pX} - 1\right). \tag{6.59}$$

This shows that the polarization in the x–z plane, whether linear, circular, or elliptic, depends on the source frequency and the plasma parameters. Moreover, by using Equations 6.52 and 6.49, the relation between $\mathbf{J}$ and $\mathbf{E}$ can be written as

$$\mathbf{J} = \varepsilon_0 \overline{\sigma} \cdot \mathbf{E}, \tag{6.60}$$

$$\begin{bmatrix} J_x \\ J_z \end{bmatrix} = j\omega\varepsilon_0 \begin{bmatrix} \left(\varepsilon_{pX} - 1\right) & 0 \\ 0 & -1 \end{bmatrix}\begin{bmatrix} E_x \\ E_z \end{bmatrix}. \tag{6.61}$$

where $\overline{\sigma}$ is the conductivity tensor for the case under consideration. The dielectric modeling of the plasma may be deduced from

$$\mathbf{D} = \varepsilon_0 \overline{\mathbf{K}} \cdot \mathbf{E}, \tag{6.62}$$

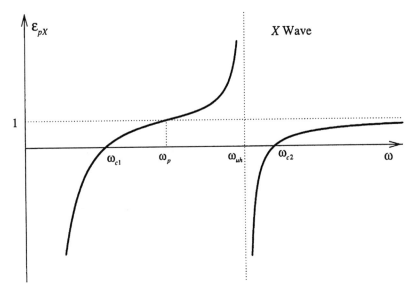

FIGURE 6.5 Dielectric constant for an X wave.

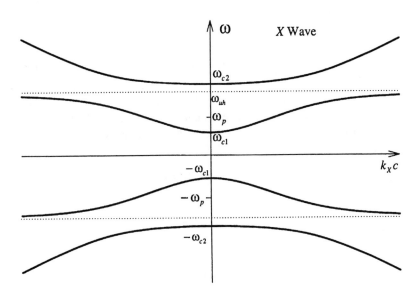

FIGURE 6.6 ω–k diagram for an X wave.

where the dielectric tensor $\overline{\mathbf{K}}$ is related to $\overline{\sigma}$ by

$$\overline{\mathbf{K}} = \overline{\mathbf{I}} + \frac{\overline{\sigma}}{j\omega\varepsilon_0}, \qquad (6.63)$$

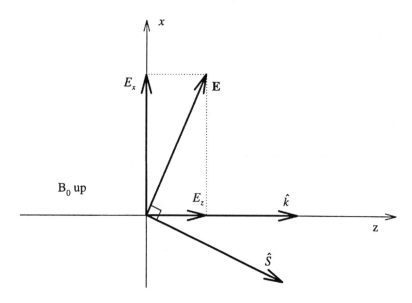

FIGURE 6.7 Geometrical sketch of the directions of various components of the X wave.

which gives, for this case,

$$\begin{bmatrix} D_x \\ D_z \end{bmatrix} = \begin{bmatrix} \varepsilon_{pX} & 0 \\ 0 & 0 \end{bmatrix} \begin{bmatrix} E_x \\ E_z \end{bmatrix}. \tag{6.64}$$

This shows that $D_z = 0$, while $E_z \neq 0$. For the X wave **D**, **H**, and the direction of propagation are mutually orthogonal. Although **E** and **H** are orthogonal, **E** and $\hat{\mathbf{k}}$ (the unit vector in the direction of propagation) are not orthogonal. We can summarize by stating

$$\hat{\mathbf{E}} \times \hat{\mathbf{H}} \neq \hat{\mathbf{k}}, \hat{\mathbf{D}} \times \hat{\mathbf{H}} = \hat{\mathbf{k}}, \tag{6.65}$$

$$E_x = \eta_{pX} H_y = \frac{\eta_0}{\sqrt{\varepsilon_{pX}}} = \frac{\eta_0}{n_{pX}}, \tag{6.66}$$

$$\mathbf{E} \cdot \hat{\mathbf{k}} \neq 0, \mathbf{D} \cdot \hat{\mathbf{k}} = 0. \tag{6.67}$$

In the above, n_{pX} is the refractive index. Figure 6.7 is a geometrical sketch of the directions of various components of the X wave. The direction of power flow is given by the Poynting vector $\mathbf{S} = \mathbf{E} \times \mathbf{H}$, which in this case is not in the direction of phase propagation. The result that $D_z = 0$ for the plane wave comes from the general equation

$$\nabla \cdot \mathbf{D} = 0 \tag{6.68}$$

in a sourceless medium. In an anisotropic medium, it does not necessarily follow that the divergence of the electric field is also zero because of the tensorial nature of the constitutive relation. For example, from the expression

$$D_z = \varepsilon_0 \left[K_{zx} E_x + K_{zy} E_y + K_{zz} E_z \right] \tag{6.69}$$

D_z can be zero without E_z being zero.

6.6 DIELECTRIC TENSOR OF A LOSSY MAGNETOPLASMA MEDIUM

The constitutive relation for a lossy plasma with collision frequency v (rad/s) is obtained by modifying Equation 6.2 further:

$$\frac{d\mathbf{J}}{dt} + v\mathbf{J} = \varepsilon_0 \omega_p^2(\mathbf{r},t)\mathbf{E} - \mathbf{J} \times \boldsymbol{\omega}_b. \tag{6.70}$$

By assuming time harmonic variation exp $(j\omega t)$ and taking the z-axis as the direction of the static magnetic field,

$$\boldsymbol{\omega}_b = \hat{z}\omega_b, \tag{6.71}$$

Equation 6.70 can be written as

$$(v + j\omega) \begin{bmatrix} J_x \\ J_y \end{bmatrix} = \varepsilon_0 \omega_p^2 \begin{bmatrix} E_x \\ E_y \end{bmatrix} - \omega_b \begin{bmatrix} J_y \\ -J_x \end{bmatrix} \tag{6.72}$$

and

$$J_z = \frac{\varepsilon_0 \omega_p^2}{v + j\omega} E_z. \tag{6.73}$$

Equation 6.72 can be written as

$$\begin{bmatrix} v + j\omega & \omega_b \\ -\omega_b & v + j\omega \end{bmatrix} \begin{bmatrix} J_x \\ J_y \end{bmatrix} = \varepsilon_0 \omega_p^2 \begin{bmatrix} E_x \\ E_y \end{bmatrix}. \tag{6.74}$$

$$\begin{bmatrix} J_x \\ J_y \end{bmatrix} = \varepsilon_0 \omega_p^2 \begin{bmatrix} v + j\omega & \omega_b \\ -\omega_b & v + j\omega \end{bmatrix}^{-1} \begin{bmatrix} E_x \\ E_y \end{bmatrix}. \tag{6.75}$$

By combining Equation 6.75 with Equation 6.73,

$$
\begin{bmatrix} J_x \\ J_y \\ J_z \end{bmatrix} = \overline{\sigma} \cdot \mathbf{E}, \tag{6.76}
$$

where

$$
\overline{\sigma} = \begin{bmatrix} \sigma_\perp & -\sigma_H & 0 \\ \sigma_H & \sigma_\perp & 0 \\ 0 & 0 & \sigma_\| \end{bmatrix}. \tag{6.77}
$$

In the above $\sigma_\perp$, σ_H, and $\sigma_\|$ are called perpendicular conductivity, Hall conductivity, and parallel conductivity, respectively. From Equation 6.63 the dielectric tensor $\overline{K}$ can now be obtained:

$$
\overline{K} = \begin{bmatrix} \varepsilon_{r\perp} & -j\varepsilon_{rH} & 0 \\ j\varepsilon_{rH} & \varepsilon_{r\perp} & 0 \\ 0 & 0 & \varepsilon_{r\|} \end{bmatrix}, \tag{6.78}
$$

where

$$
\varepsilon_{r\perp} = 1 - \frac{\left(\omega_p/\omega\right)^2 \left(1 - jv/\omega\right)}{\left(1 - j\dfrac{v}{\omega}\right)^2 - \left(\omega_b/\omega\right)^2}, \tag{6.79}
$$

$$
\varepsilon_{rH} = \frac{\left(\omega_p/\omega\right)^2 \left(\omega_b/\omega\right)}{\left(1 - jv/\omega\right)^2 - \left(\omega_b/\omega\right)^2}, \tag{6.80}
$$

$$
\varepsilon_{r\|} = 1 - \frac{\left(\omega_b/\omega\right)^2}{1 - jv/\omega}. \tag{6.81}
$$

6.7   PERIODIC LAYERS OF MAGNETOPLASMA

Layered semiconductor–dielectric periodic structures in the presence of a dc magnetic field can be modeled as periodic layers of magnetoplasma. Electromagnetic waves propagating in such structures can be investigated by using the dielectric tensor derived in Section 6.6. The solution is obtained by extending the theory of Section 2.7 to the case where the plasma layer is magnetized and, hence, has the anisotropy complexity. Brazis and Safonova[12-14] investigated the resonance and absorption bands of such structures.

6.8 SURFACE MAGNETOPLASMONS

The theory of Section 2.8 may by extended to the propagation of surface waves at a semiconductor–dielectric interface in the presence of a static magnetic field. The semiconductor in a dc magnetic field can be modeled as an anisotropic dielectric tensor. Wallis and Quinn[15] discuss at length the properties of surface magnetoplasmons on semiconductors.

6.9 SURFACE MAGNETOPLASMONS IN PERIODIC MEDIA

References 16 and 17 discuss the propagation of surface magnetoplasmons in truncated superlattices. This model is a combination of the models used in Sections 6.7 and 6.8.

REFERENCES

1. Heald, M. A. and Wharton, C. B., *Plasma Diagnostics with Microwaves*, Wiley, New York, 1965.
2. Booker, H. G., *Cold Plasma Waves*, Kluwer, Hingham, MA, 1984.
3. Swanson, D. G., *Plasma Waves*, Academic Press, New York, 1989.
4. Steele, M. C., *Wave Interactions in Solid State Plasmas*, McGraw-Hill, New York, 1969.
5. Bowers, R., Legendy, C., and Rose, F., Oscillatory galvanometric effect in metallic sodium, *Phys. Rev. Lett.*, 7, 339, 1961.
6. Aigrain, P. R., in *Proceedings of the International Conference on Semiconductor Physics*, Czechoslovak Academy Sciences, Prague, 224, 1961.
7. Miyamoto, K., *Plasma Physics for Nuclear Fusion*, MIT Press, Cambridge, MA, 1976.
8. Solymar, L. and Walsh, D., *Lectures on the Electrical Properties of Materials*, 5th ed., Oxford University Press, Oxford, 1993.
9. Madala, S. R. V. and Kalluri, D. K., Longitudinal propagation of low frequency waves in a switched magnetoplasma medium, *Radio Sci.*, 28, 121, 1993.
10. Dimitrijevic, M. M. and Stanic, B. V., EMW transformation in suddenly created two-component magnetized plasma, *IEEE Trans. Plasma Sci.*, 23, 422, 1995.
11. Madala, S. R. V., Frequency Shifting of Low Frequency Electromagnetic Waves Using Magnetoplasmas, Ph.D. Thesis, University Massachusetts, Lowell, 1993.
12. Brazis, R. S. and Safonova, L. S., Resonance and absorption band in the classical magnetoactive semiconductor-insulator superlattice, *Int. J. Infrared Millimeter Waves*, 8, 449, 1987.
13. Brazis, R. S. and Safonova, L. S., Electromagnetic waves in layered semiconductor-dielectric periodic structures in dc magnetic fields, *Proc. SPIE*, 1029, 74, 1988.

14. Brazis, R. S. and Safonova, L. S., In-plane propagation of millimeter waves in periodic magnetoactive semiconductor structures, *Int. J. Infrared Millimeter Waves,* 18, 1575, 1997.
15. Wallis, R. F., Surface magnetoplasmons on semiconductors, in *Electromagnetic Surface Modes*, Boardman, A. D., Ed., John Wiley, New York, 1982, chap. 2.
16. Wallis, R. F., Szenics, R., Quihw, J. J., and Giuliani, G. F., Theory of surface magnetoplasmon polaritons in truncated superlattices, *Phys. Rev. B*, 36, 1218, 1987.
17. Wallis, R. F. and Quinn, J. J., Surface magnetoplasmon polaritons in truncated superlattices, *Phys. Rev. B*, 38, 4205, 1988.

7 Switched Magnetoplasma Medium

7.1 INTRODUCTION

The effect of converting free space into an anisotropic medium by creation of a plasma medium in the presence of a static magnetic field is considered in this chapter. In Section 7.2 the higher-order differential equations for fields in a time-varying magnetoplasma medium are developed. These equations are the basis of adiabatic analysis discussed in Chapter 9.

The main effect of switching the medium is the splitting of the original wave (incident wave) into new waves whose frequencies are different from the frequency of the incident wave. The strength and the direction of the static magnetic field affect the number of waves generated, their frequencies, damping rates, and their power densities. These aspects for L, R, O, and X waves are discussed in Sections 7.3 through 7.9.

7.2 ONE-DIMENSIONAL EQUATIONS: LONGITUDINAL PROPAGATION

First, the propagation of R and L plane waves in a time-varying space-invariant unbounded magnetoplasma medium is considered, i.e.,

$$\omega_p^2 = \omega_p^2(t), \tag{7.1}$$

$$\omega_b = \omega_b(t), \tag{7.2}$$

$$F(z,t) = F(t)\exp(-jkz), \tag{7.3}$$

$$\frac{\partial}{\partial z} = -jk. \tag{7.4}$$

The relevant equations are obtained from Section 6.3:

$$\mu_0 \frac{\partial H}{\partial t} = jkE, \tag{7.5}$$

$$\varepsilon_0 \frac{\partial E}{\partial t} = jkH - J, \tag{7.6}$$

$$\frac{dJ}{dt} = \varepsilon_0 \omega_p^2(t)E \pm j\omega_b(t)J. \tag{7.7}$$

From Equation 6.20 or from Equations 7.5 through 7.7, the equation for H can be obtained:

$$\frac{d^3H}{dt^3} \mp j\omega_b(t)\frac{d^2H}{dt^2} + \left[k^2c^2 + \omega_p^2(t)\right]\frac{dH}{dt} \mp j\omega_b(t)k^2c^2H = 0. \tag{7.8}$$

The equation for E from Equation 6.18 turns out to be more complicated than the equation for H. However, if ω_p^2 alone is a function of time and ω_b is a constant, a third-order differential equation for E involving the first derivative of $\omega_p^2(t)$ is obtained:

$$\frac{d^3E}{dt^3} \mp j\omega_b\frac{d^2E}{dt^2} + \left[k^2c^2 + \omega_p^2(t)\right]\frac{dE}{dt} + \left(g \mp j\omega_b k^2c^2\right)E = 0, \tag{7.9}$$

where

$$g = \frac{d\omega_p^2(t)}{dt}. \tag{7.10}$$

No such simplification is possible if ω_p^2 is a constant, but ω_b varies with time. In such a case it is still possible to write a third-order equation for J. However, it includes the first and second derivatives of $\omega_b(t)$ denoted by $\dot{\omega}_b$ and $\ddot{\omega}_b$, respectively:

$$\frac{d^3J}{dt^3} \mp j\omega_b(t)\frac{d^2J}{dt^2} + \left[k^2c^2 + \omega_p^2(t) \mp 2j\dot{\omega}_b(t)\right]\frac{dJ}{dt} \mp j\left(k^2c^2\omega_b(t) + \ddot{\omega}_b(t)\right)J = 0. \tag{7.11}$$

The equation for the isotropic case should be obtained by substituting $\omega_b = 0$ in the above equations; Equation 7.8 indeed reduces to Equation 3.13. Equation 7.9 remains third order in E even if $\omega_b = 0$ is substituted, whereas the equation for the isotropic case (Equation 3.11) is of second order. However, Equation 7.9 is obtained by differentiating Equation 3.11 with respect to t.

7.3 SUDDEN CREATION: LONGITUDINAL PROPAGATION

Next the problem of sudden creation of the plasma in the presence of a static magnetic field in the z-direction is solved. The geometry of the problem is given in Figure 7.1. It will be assumed that the source wave is a circularly polarized wave propagating in the z-direction. Solution for R and L waves can be done simultaneously.

This problem will be similar to that of Section 3.2 if it is assumed that the source wave is an R or L wave. The Laplace transform technique will again be used. The

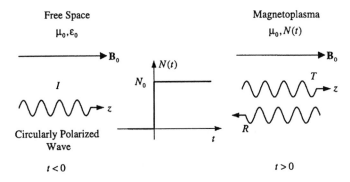

FIGURE 7.1 Geometry of the problem. Magnetoplasma medium is switched on at $t = 0$.

state variable technique could be used, but to add to variety, the higher-order differential equation for H rather than the state variable equations will be solved. By assuming that the source wave medium is free space and its frequency is $\omega_0 = kc$,

$$\frac{d^3 H}{dt^3} \mp j\omega_b \frac{d^2 H}{dt^2} + \left[\omega_0^2 + \omega_p^2\right]\frac{dH}{dt} \mp j\omega_b \omega_0^2 H = 0. \tag{7.12}$$

The initial conditions on E, H, and J given by Equations 3.25 through 3.27 can be converted into the initial conditions on H and its derivatives at the origin:

$$H(0) = H_0, \tag{7.13}$$

$$\dot{H}(0) = j\omega_0 H_0, \tag{7.14}$$

$$\ddot{H}(0) = -\omega_0^2 H_0. \tag{7.15}$$

After some algebra,

$$\frac{H(s)}{H_0} = \frac{\left(s \mp j\omega_b\right)\left(s + j\omega_0\right) + \omega_p^2}{\left(s - j\omega_1\right)\left(s - j\omega_2\right)\left(s - j\omega_3\right)}, \tag{7.16}$$

where ω_1, ω_2, and ω_3 are the roots of the polynomial equation:

$$\omega^3 \mp \omega_b \omega^2 - \left(\omega_0^2 + \omega_p^2\right)\omega \pm \omega_0^2 \omega_b = 0. \tag{7.17}$$

By evaluating the residues at the poles, $H(s)$ can be inverted. The solution for $H(t)$ can be expressed in a compact form

$$\frac{H(t)}{H_0} = \sum_{n=1}^{3} \frac{H_n}{H_0} \exp\left(j\omega_n t\right), \tag{7.18}$$

where

$$\frac{H_n}{H_0} = \frac{\left(\omega_n \mp \omega_b\right)\left(\omega_n + \omega_0\right) - \omega_p^2}{\displaystyle\prod_{m=1,m\neq n}^{3} \left(\omega_n - \omega_m\right)}. \tag{7.19}$$

The electric field of each of the waves can now easily be obtained by noting from Equation 7.5

$$\frac{E_n}{H_n} = \eta_0 \frac{\omega_n}{\omega_0}. \tag{7.20}$$

Of course, the electric field can also be obtained by solving the differential Equation 7.9. Care must be taken in interpreting g for the suddenly created plasma. The $\omega_p^2(t)$ profile is expressed in terms of a step profile

$$\omega_p^2(t) = \omega_p^2 u(t), \tag{7.21}$$

and $g(t)$, its derivative, can be expressed in terms of an impulse function $\delta(t)$:

$$g(t) = \omega_p^2 \delta(t). \tag{7.22}$$

Equation 7.9 becomes

$$\frac{d^3 E}{dt^3} \mp j\omega_b \frac{d^2 E}{dt^2} + \left[\omega_0^2 + \omega_p^2\right]\frac{dE}{dt} + \left[\omega_p^2 \delta(t) \mp j\omega_b \omega_0^2\right]E = 0. \tag{7.23}$$

Equation 7.23 can now be solved subject to the initial condition

$$E(0) = E_0, \tag{7.24}$$

$$\dot{E}(0) = j\omega_0 E_0, \tag{7.25}$$

$$\ddot{E}(0) = -\omega_0^2 E_0. \tag{7.26}$$

By noting

$$\pounds\{\delta(t)E(t)\} = \int_0^\infty \delta(t)\exp(-st)E(t)\,dt = E(0) = E_0, \tag{7.27}$$

the following are obtained:

$$\frac{E(s)}{E_0} = \frac{\left(s \mp j\omega_b\right)\left(s + j\omega_0\right)}{\left(s - j\omega_1\right)\left(s - j\omega_2\right)\left(s - j\omega_3\right)} \tag{7.28}$$

and

$$\frac{E_n}{E_0} = \frac{\left(\omega_n \mp \omega_b\right)\left(\omega_n + \omega_0\right)}{\displaystyle\prod_{m=1,m\neq n}^{3}\left(\omega_n - \omega_m\right)}. \tag{7.29}$$

After a little algebra, it can be shown that E_n obtained from Equation 7.19 is the same as that given by Equation 7.29. Moreover, it can be shown that

$$\frac{E_n}{H_n} = \sqrt{\frac{\mu_0}{\varepsilon_0 \varepsilon_{pn}}} = \frac{\eta_0}{n_{pn}}, \tag{7.30}$$

where

$$\varepsilon_{pn} = 1 - \frac{\omega_p^2}{\omega_n\left(\omega_n \mp \omega_b\right)}. \tag{7.31}$$

Here, ε_{pn} is the dielectric constant and n_{pn} is the refractive index at the frequency ω_n. The product of Equation 7.29 with Equation 7.19 gives the ratio of the power in the nth wave to the power in the incident wave, called S_n. In general,

$$S_n = \frac{E_n}{E_0}\frac{H_n}{H_0}. \tag{7.32}$$

A negative value for S_n indicates that the nth wave is a reflected wave, further confirmed by the value of ω_j. The ω–k diagram discussed in the next paragraph confirmed that the group velocity of these waves has the same algebraic sign as the phase velocity, thus ruling out any backward waves (see Reference 3, p. 257). The sum of the absolute values of S_n gives the ratio of total power in the created waves to the power in the incident wave:

$$S = \sum_{n=1}^{3}\left|S_n\right|. \tag{7.33}$$

The effect of switching the magnetoplasma medium is the creation of three waves with new frequencies (frequencies other than the incident wave frequency) given by Equation 7.17. These new frequencies are functions of ω_0, ω_b, and ω_p. Their variation is conveniently studied with the use of normalized variables defined in Equations 7.34 to 7.37.

$$\Omega = \frac{\omega}{\omega_p}, \tag{7.34}$$

$$\Omega_0 = \frac{\omega_0}{\omega_p}, \tag{7.35}$$

$$\Omega_n = \frac{\omega_n}{\omega_p}, n = 1, 2, 3, \tag{7.36}$$

$$\Omega_b = \frac{\omega_b}{\omega_p}. \tag{7.37}$$

In terms of these normalized variables, Equation 7.17 becomes

$$\Omega^3 \mp \Omega_b \Omega^2 - \left(1 + \Omega_0^2\right)\Omega \pm \Omega_0^2 \Omega_b = 0, \tag{7.38}$$

where the lower sign is to be used when the incident wave is an L wave and the upper sign when the incident wave is an R wave. Sketches of the curves Ω vs. Ω_0 with Ω_b as a parameter are given for the two cases in Figures 7.2 and 7.3, respectively. It can be noted that the curve Ω vs. Ω_0 is also the ω–k diagram.

$$\Omega_0 = \frac{\omega_0}{\omega_p} = \frac{k_0 c}{\omega_p} = \frac{kc}{\omega_p}. \tag{7.39}$$

In the above, k_0 is the wave number of the incident wave, which is also the wave number k of the newly created waves since the wave number is conserved at a time discontinuity. As expected, they are similar to the curves of Figure 6.4 and Figure 6.2, respectively. The phase velocity v_{ph} of the waves is given by

$$v_{ph} = \frac{\omega}{k} = c\frac{\Omega}{\Omega_0}. \tag{7.40}$$

The group velocity v_{gr} is given by Equation 7.41:

$$v_{gr} = c\frac{d\Omega}{d\Omega_0}. \tag{7.41}$$

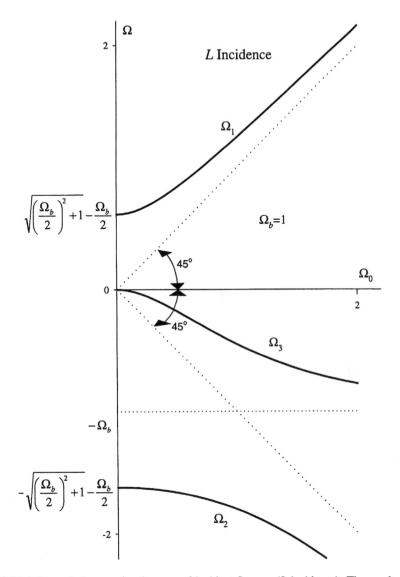

FIGURE 7.2 ω–k diagram for the case of incident L wave (L incidence). The results are presented in terms of normalized variables. The horizontal axis is $\Omega_0 = \omega_0/\omega_p$, where ω_p is the plasma frequency of the switched medium. Ω_0 is also proportional to the wave number of the created waves. The vertical axis is $\Omega = \omega/\omega_p$, where ω is the frequency of the created waves. The branch marked Ω_n describes the nth created wave. The parameter is $\Omega_b = \omega_b/\omega_p$, where ω_b is the gyrofrequency of the switched medium.

The branch marked Ω_1 in Figure 7.2 is the dispersion curve for the first L wave. This is a transmitted wave. The asymptote to this branch at 45° shows that the phase and group velocities of this wave approach those of the incident wave as $\Omega_0 = \omega_0/\omega_p$ becomes large. When the frequency of the incident wave is much larger than the plasma frequency, the plasma has little effect on the incident wave. The branch

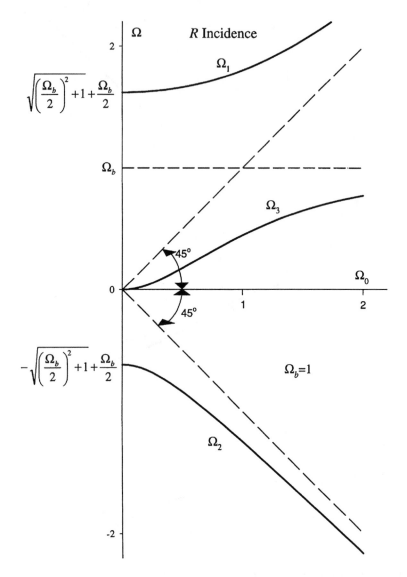

FIGURE 7.3 ω–k diagram for the case of incident R wave (R incidence).

marked Ω_2 describes the reflected wave whose asymptote is at $-45°$. The branch marked Ω_3 describes the third wave, which is a reflected wave. Its asymptote is a horizontal line $\Omega = -\Omega_b$. Its phase and group velocities are negative. The third wave disappears in the isotropic case ($\Omega_b = 0$).

Figure 7.3 describes the dispersion for the three waves when the incident wave is the R wave. In this case, the third wave is a transmitted wave with an asymptote $\Omega = \Omega_b$. The following differences between the case studied here and the isotropic case ($\Omega_b = 0$) reported in the literature[4,5] can be noted.

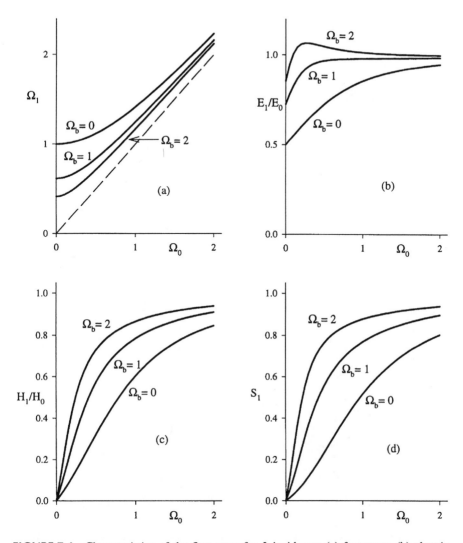

FIGURE 7.4 Characteristics of the first wave for L incidence: (a) frequency, (b) electric field, (c) magnetic field, (d) power density.

1. The first and the second waves do not have the same frequencies.
2. The third wave does not exist in the isotropic case. Instead, a static but spatially varying magnetic field exists.[4] The same result, i.e., $\omega_3 = 0$, E_3/E_0 = 0, and $H_3/H_0 = \omega_p^2/(\omega_0^2 + \omega_p^2)$ is obtained, when $\omega_b = 0$ is substituted in Equations 7.17, 7.29, and 7.19.

7.4 NUMERICAL RESULTS: LONGITUDINAL PROPAGATION

In this section, the characteristics of the three waves mentioned above are studied in detail. The horizontal axis is the normalized frequency (normalized with respect

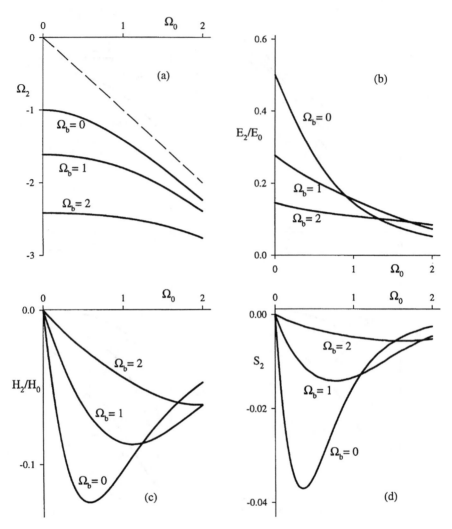

FIGURE 7.5 Characteristics of the second wave for L incidence: (a) frequency, (b) electric field, (c) magnetic field, (d) power density.

to the plasma frequency) of the incident wave. The normalized gyrofrequency (Ω_b) is used as a parameter. The case of incident L wave (for brevity this case will be referred to as L incidence) is shown in Figures 7.4 through 7.7 and the case of incident R wave (R incidence) in Figures 7.8 through 7.11.

Figure 7.4 describes the first wave for L incidence, which is a transmitted wave. In Figure 7.4a the variation of the normalized frequency of the first wave is shown. The isotropic case is described by the curve $\Omega_b = 0$. For strong static magnetic fields (high Ω_b) and low values of the incident wave frequencies, the frequency of the first wave can be made smaller than the plasma frequency. In Figure 7.4b the electric field is shown; it shows a peak value of 1.07 at about $\Omega_0 = 0.27$ for $\Omega_b = 2$. For a

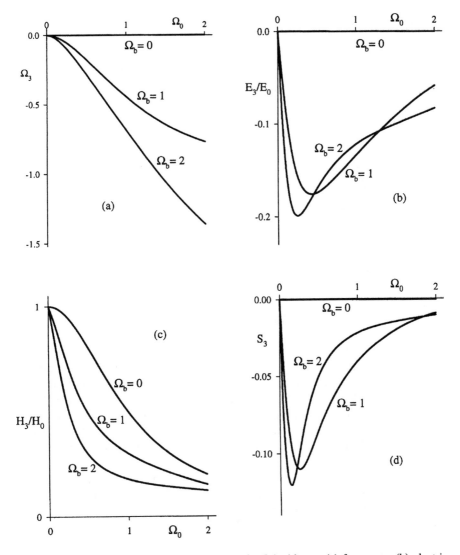

FIGURE 7.6 Characteristics of the third wave for L incidence: (a) frequency, (b) electric field, (c) magnetic field, (d) power density.

very large value of $\Omega_b(\Omega_b = 10)$, the peak value is about 1.2 occurring at $\Omega_0 = 0.05$ (not shown in the figure). In Figure 7.4c the wave magnetic field and in Figure 7.4d the time-averaged power density are shown. The static magnetic field improves the power level of this wave. For a large value of Ω_0 the frequency of the first wave asymptotically approaches the incident wave frequency. Plasma has very little effect.

Figure 7.5 describes the second wave, which is a reflected wave. Its frequency is always greater than the incident wave frequency. Its power level peaks to 4% at about $\Omega_0 = 0.35$ and $\Omega_b = 0$. The static magnetic field reduces the power level of this reflected wave.

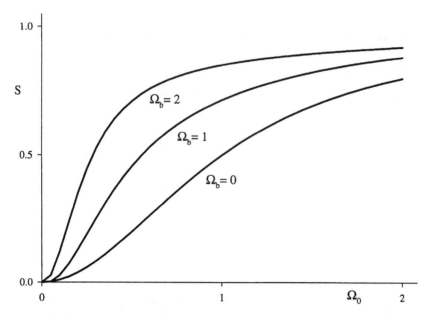

FIGURE 7.7 Total power density of the created waves for L incidence.

Figure 7.6 describes the third wave, which in this case (L incidence) is a reflected wave. This wave is due to the presence of the static magnetic field and is strongly influenced by it. The frequency of this wave is less than Ω_b. The power level peaks to about 12% (at $\Omega_0 = 0.1$ for $\Omega_b = 0.1$), and the peaking point is influenced by Ω_b. Figure 7.7 shows the ratio of the total power in the newly created waves to the power in the incident wave. It is always less than 1. Some of the energy of the incident wave gets transferred as kinetic energy of the electrons during the creation of the plasma.[4]

Figures 7.8 through 7.11 describe R incidence. A static magnetic field causes a drop in the power level of the transmitted wave (Figure 7.8d) and causes an increase in the peak power level of the reflected wave (Figure 7.9d). This behavior is in contrast to the behavior for L incidence. (Compare Figure 7.4d with 7.8d, and Figure 7.5d with 7.9d.)

The third wave, for R incidence, is a transmitted wave whose frequency is again strongly controlled by the static magnetic field. The power of this wave (Figure 7.10d) peaks (about 70% for $\Omega_b = 2$, at $\Omega_0 = 0.6$) at an optimum Ω_0. For a very large value of Ω_b ($\Omega_b = 12$), the peak value of the power is almost 100%. In the range of Ω_0 shown in Figure 7.11a, $\Omega_3 \approx \Omega_0$ for $\Omega_b = 12$; the graph of Ω_3 vs. Ω_0 is a 45° line. In this range, most of the incident power is carried by the third wave. At higher incident wave frequencies ($\Omega_0 > 12$), the Ω_3 vs. Ω_0 curve flattens and approaches the horizontal asymptote $\Omega_3 = \Omega_b$ (not shown in the figure). In this higher range of Ω_0, most of the incident power will be carried by the first wave.

A physical interpretation of the three waves may be given in the following way: the electric field of the incident wave accelerates the electrons in the newly created magnetoplasma, which in turn radiates new waves. The frequencies (ω) of the new

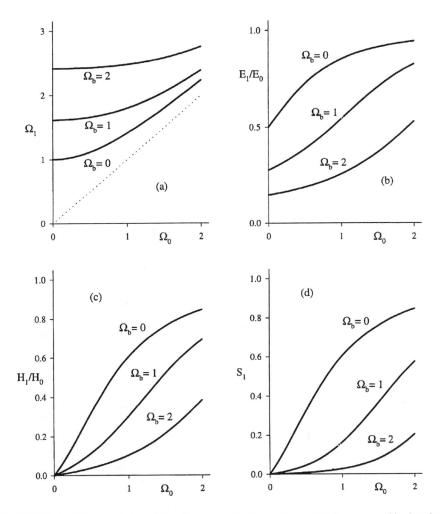

FIGURE 7.8 Characteristics of the first wave for R incidence: (a) frequency, (b) electric field, (c) magnetic field, (d) power density.

waves are constrained by the requirements that the wave number be conserved and the refractive index (n) be that applicable to longitudinal propagation in magnetoplasma.

$$\omega = \beta v_{ph} = k v_{ph} = \frac{\omega_0}{n},\qquad(7.42)$$

where

$$n = \sqrt{1 - \frac{\omega_p^2}{\omega(\omega \mp \omega_b)}}.\qquad(7.43)$$

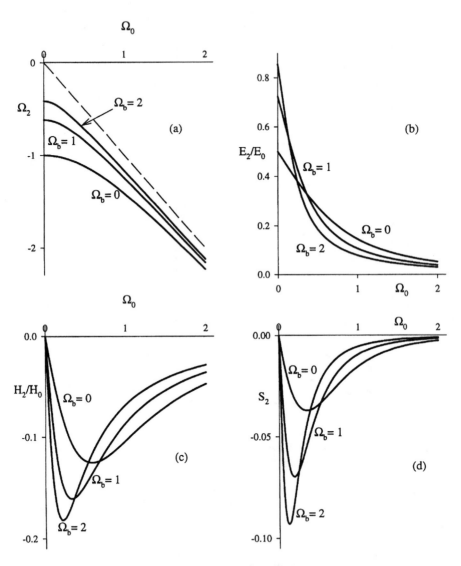

FIGURE 7.9 Characteristics of the second wave for R incidence: (a) frequency, (b) electric field, (c) magnetic field, (d) power density.

Equation 7.42, when simplified, reduces to Equation 7.17.

The propagation transverse to the imposed magnetic field results in ordinary or extraordinary waves (X wave). The ordinary wave case is the same as the isotropic case. The X wave case is discussed in Section 7.6.

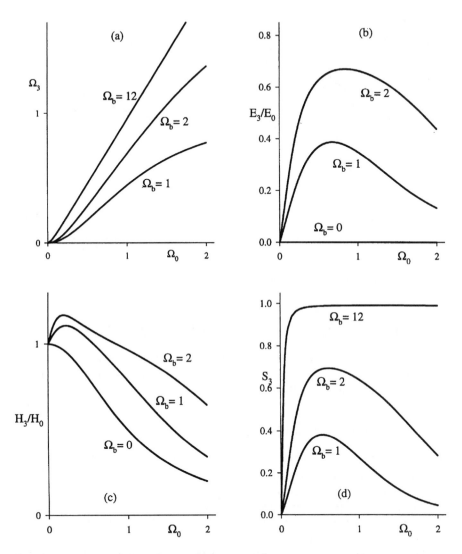

FIGURE 7.10 Characteristics of the third wave for R incidence: (a) frequency, (b) electric field, (c) magnetic field, (d) power density.

7.5 DAMPING RATES: LONGITUDINAL PROPAGATION

The analysis thus far has ignored the collision frequency ν. Assuming a low-loss plasma medium where $\nu/\omega_p \ll 1$, Reference 6 gives expressions for the damping time constant and attenuation length for longitudinal propagation of the B wave in a magnetoplasma. The reprint of Reference 6 is given in this book as Appendix B.

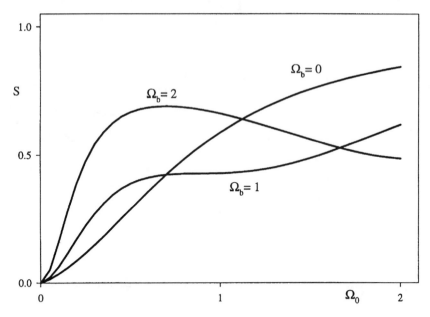

FIGURE 7.11   Total power density of the created waves for R incidence.

7.6   SUDDEN CREATION: TRANSVERSE PROPAGATION, X WAVE

The complete theory of "wave propagation in a switched magnetoplasma medium: transverse propagation" is discussed in Reference 7, and the reprint of this paper is given in this book as Appendix C.

7.7   ADDITIONAL NUMERICAL RESULTS

It has been seen that the strength and the direction of the static magnetic field affect the number of new waves created, their frequencies, and the power density of these waves. Some of the new waves are transmitted waves (waves propagating in the same direction as the incident wave), and some are reflected waves (waves propagating in the opposite direction to that of the incident wave). Reflected waves tend to have less power density.

The shift ratio and the efficiency of the frequency-shifting operation may be controlled by the strength and the direction of the static magnetic field. It is one of the two aspects that is discussed in this section.

In Figure 7.12 the results are presented for the $L1$ wave. It has a frequency Ω_1 $> \Omega_0$ and is a transmitted wave with considerable power in it. Figure 7.12a and b shows the variation of Ω_1 and S_1 with Ω_0 for various values of Ω_b. Figure 7.12c is constructed from Figure 7.12a and b and shows the frequency upshift ratio Ω_1/Ω_0 vs. Ω_0. The solid-line and the broken-line curves have percentage power and the electron gyrofrequency (Ω_b) as parameters, respectively. The beneficial effect of the

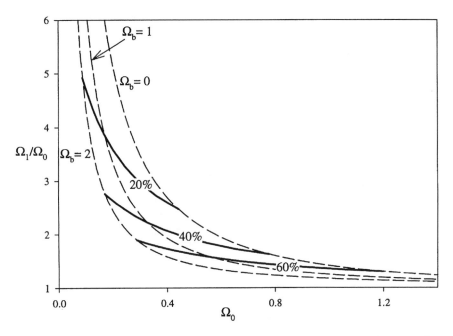

FIGURE 7.12 Frequency-upshifting of $L1$ wave. Frequency-shift ratio as a function of the incident wave frequency (ω_0). The solid-line and the broken-line curves have S_1 expressed in percent and the electron gyrofrequency (ω_b) as parameters, respectively. All frequency variables are normalized with respect to the plasma frequency (ω_p): $\Omega_b = \omega_b/\omega_p$, $\Omega_0 = \omega_0/\omega_p$, $\Omega_1 = \omega_1/\omega_p$. The power density is normalized with respect to that of the source wave.

strength of the static magnetic field in increasing the power for a given shift ratio is evident from the Figure 7.12.

In Figures 7.13 and 7.14 the cases of $R1$ and $R3$ waves are considered, respectively. $R1$ is upshifted, whereas $R3$ is downshifted.

Figures 7.15 and 7.16 show the frequency shifts of the $X1$ and $X2$ waves. The $X1$ wave is upshifted, and the $X2$ wave is upshifted for $\Omega_0 < 1$ and downshifted for $\Omega_0 > 1$.

The waves generated in the switched medium are damped as they travel in the medium if it is lossy. The formulas for the skin depth (d_{pn}) and the damping time constant (t_{pn}) were derived earlier. See Appendexes B and C.

Table 7.1 is prepared from the results presented in the figures and lists the parameters of the switched magnetoplasma medium for a given shift ratio. From the shift ratio $f_n/f_0 = \Omega_n/\Omega_0$, the variables Ω_0 and S_n/S_0 are found from the figures for $\Omega_b = 0$, 1, and 2. For an assumed value of f_0, Ω_b and ω_p are determined and these are converted into B_0 and N_0. It can be noted that $R3$ and $X2$ waves do not exist for the isotropic case ($\Omega_b = 0$). The numbers for B_0 and N_0 in Table 7.1 are computed assuming $f_0 = 10$ GHz. These numbers can be obtained for other values of f_0 by noting that N_0 scales as f_0^2 and B_0 scales as f_0. The damping constants are calculated assuming a low-loss plasma with $\nu/\omega_p = 0.01$, where ν is the collision frequency. The last column in Table 7.1 lists the answer to the question of whether or not the

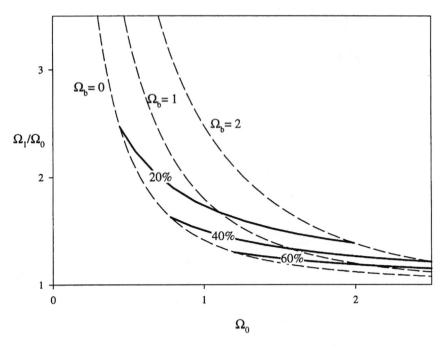

FIGURE 7.13 Frequency-upshifting of $R1$ wave. The meaning of the symbols is given in Figure 7.12.

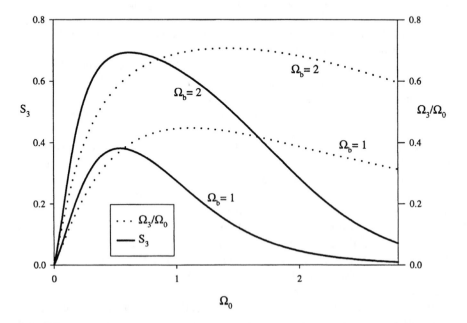

FIGURE 7.14 Frequency-downshifting of $R3$ wave. The meaning of the symbols is given in Figure 7.12.

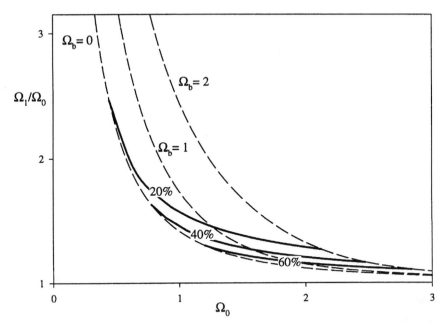

FIGURE 7.15 Frequency-upshifting of $X1$ wave. The meaning of the symbols is given in Figure 7.12.

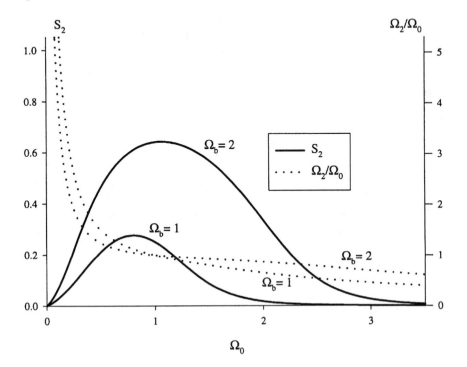

FIGURE 7.16 Frequency-shifting of $X2$ wave. The meaning of the symbols is given in Figure 7.12.

TABLE 7.1

Parameters of the Switched Magnetoplasma Medium for a Given Shift Ratio f_n/f_0

Line No.	Shift Ratio	Wave	B_0, Wb/m³	N_0, 10^{18}/m³	δ_{pn}, m	t_{pn}, ns	S_n/S_0, %	In Stop Band?
1	2.3	O	0	5.36	0.24	1.9	22	Yes
2	2.3[a]	$L1$	1.15	12.8	0.38	2.25	36	Yes
3	2.3[b]	$L1$	3.25	25.5	0.57	2.82	48	Yes
4	2.3[a]	$X2$	1.19	13.7	N.D.[c]	0.74	6	Yes
5	2.3[b]	$X2$	3.57	30.9	0.20	0.88	25	Yes
6	1.5	O	0	1.52	1.05	5.12	47	Yes
7	1.5[a]	$L1$	0.59	3.32	1.42	6.4	62	Yes
8	1.5[b]	$L1$	1.59	6.1	2.05	8.07	73	No
9	1.5[a]	$R1$	0.28	0.76	0.68	4.63	26	Yes
10	1.5[b]	$R1$	0.40	0.38	0.56	4.69	14	Yes
11	1.5[a]	$X1$	0.32	0.98	0.36	3.4	11	No
12	1.5[b]	$X1$	0.43	0.45	0.26	3.84	4	No
13	1.5[a]	$X2$	0.75	5.47	0.45	3.42	17	Yes
14	1.5[b]	$X2$	1.79	7.72	1.05	6.35	41	Yes
15	0.7[b]	$X2$	0.20	0.10	N.D.[c]	6.41	1	No
16	0.6[b]	$R3$	0.26	0.16	0.55	6.28	7.5	No
17	0.6[b]	$R3$	1.28	3.94	2.47	10.3	69	No
18	0.4[a]	$R3$	0.19	0.34	0.25	4.4	5.5	No
19	0.4[a]	$R3$	0.58	3.21	0.84	5.2	38	No
20	0.4[b]	$R3$	2.75	18.26	2.23	10.48	55	No

Note: Incident wave frequency: $f_0 = 10$ GHz; normalized collision frequency: $\nu/\omega_p = 0.01$.

[a] $\Omega_B = 1$.

[b] $\Omega_B = 2$

[c] Not determined.

source frequency ω_0 falls in the stop band of the particular mode. An incident wave whose frequency ω_0 falls in the stop band is totally reflected by the space boundary of the medium. The free space on the forward side of the magnetoplasma medium then will not have signals at the frequency of the incident wave. From the results in the table, it appears the $L1$ wave is best suited for upshifting and the $R3$ wave for downshifting if there are no constraints on B_0 and N_0. In both cases, the strength of the magnetic field improves the power density of the frequency-shifted wave. A stronger static magnetic field requires a stronger ionization to achieve the same shift ratio. A stronger static magnetic field gives a larger time constant.

The numerical results presented so far are normalized with reference to the plasma frequency. For the convenience of experimentalists who prefer to keep the source frequency fixed and vary the plasma frequency by adjusting the vacuum pressure in the plasma vessel, the numerical results and graphs have been generated normalized with reference to the source frequency.

It has also become possible to give explicit expressions for the frequencies and the fields when $\omega_b \ll \omega_0$. These aspects are discussed in Reference 8, and a reprint of this is given in this book as Appendix D.

7.8 SUDDEN CREATION: ARBITRARY DIRECTION OF THE STATIC MAGNETIC FIELD

The switching of the plasma medium in the presence of an arbitrarily directed magnetic field creates many more frequency-shifted modes. This aspect is discussed by Dimitrijevic et al.[9]

7.9 FREQUENCY SHIFTING OF LOW-FREQUENCY WAVES

The results given so far need modification to be applicable to very low values of ω_0. The ion motion has to be introduced into the analysis, and suitable approximations corresponding to longitudinal and transverse propagation of Alfven waves can be made. These aspects are discussed in References 9 through 11.

REFERENCES

1. Kalluri, D. K., Effect of switching a magnetoplasma medium on a traveling wave: longitudinal propagation, *IEEE Trans. Antennas Propag.*, AP-37, 1638, 1989.
2. Kalluri, D. K., Frequency-shifting using magnetoplasma medium, *IEEE Trans. Plasma Sci.*, 21, 77, 1993.
3. Ramo, S., Whinnery, J. R., and Duzer, T. V., *Fields and Waves in Communication Electronics,* John Wiley, New York, 1965.
4. Jiang, C. L., Wave propagation and dipole radiation in a suddenly created plasma, *IEEE Trans. Antennas Propag.*, AP-23, 83, 1975.
5. Kalluri, D. K., On reflection from a suddenly created plasma half-space: transient solution, *IEEE Trans. Plasma Sci.*, 16, 11, 1988.
6. Kalluri, D. K. and Goteti, V. R., Damping rates of waves in a switched magnetoplasma medium: longitudinal propagation, *IEEE Trans. Plasma Sci.*, 18, 797, 1990.
7. Goteti, V. R. and Kalluri, D. K., Wave propagation in a switched magnetoplasma medium: transverse propagation, *Radio Sci.*, 25, 61, 1990.
8. Kalluri, D. K., Frequency shifting using magnetoplasma medium: flash ionization, *IEEE Trans. Plasma Sci.*, 21, 77, 1993.
9. Dimitrijevic, M. M. and Stanic, B. V., EMW transformation in suddenly created two-component magnetized plasma, *IEEE Trans. Plasma Sci.*, 23, 422, 1995.
10. Madala, S. R. V. and Kalluri, D. K., Longitudinal propagation of low frequency waves in a switched magnetoplasma medium, *Radio Sci.*, 28, 121, 1993.
11. Madala, S. R. V., Frequency Shifting of Low Frequency Electromagnetic Waves Using Magnetoplasmas, Ph.D. Thesis, University of Massachusetts, Lowell, 1993.

8 Longitudinal Propagation in a Magnetized Time-Varying Plasma

8.1 INTRODUCTION

The technique of solving this problem is discussed in Sections 3.5 through 3.9. The anisotropic case considered here is a little more difficult. In preparation for its assimilation, the reader is urged to revisit Sections 3.5 through 3.9 and Section 7.3.

Wave propagation in a transient magnetoplasma has been considered by a number of authors.[1-6] At one extreme, when the source wave period is much larger than the rise time T_r, the problem may be solved as an initial value problem using the sudden-switching approximation ($T_r = 0$).[1] The effect of switching the medium is the creation of three frequency-shifted waves. The power scattering coefficients depend on the ratio of the source frequency to the plasma frequency and gyrofrequency. The power reflection coefficient falls rapidly with the increase of T_r.

At the other extreme, the switching is considered sufficiently slow that an adiabatic analysis can be used.[2] In this approximation two of the scattering coefficients are negligible and the main forward-propagating wave is the modified source wave.

This chapter considers the wave propagation in a magnetized transient plasma for the case of a rise time comparable with the period of the source wave.[7] Geometry of the problem in the z–t plane is sketched in Figure 8.1. A right circularly polarized source wave, $R_s(\omega_1)$, of angular frequency ω_1 is propagating in an unbounded magnetized plasma medium whose parameters are the plasma frequency ω_{p1} and the gyrofrequency ω_b. The direction of wave propagation, which is also the direction of the static magnetic field, is assumed to be along the z-axis (longitudinal propagation). The plasma frequency undergoes a transient change in the interval $-T_{r1} < t < T_{r2}$, attaining its final value of ω_{p2} at $t = T_{r2}$. The transient change in the medium properties gives rise to three frequency-shifted waves, designated by R_1, R_2, and R_3.

The computation of the instantaneous frequencies and the final asymptotic frequencies of the waves is quite straightforward.[2] Enforcing the conservation of the wave number on the dispersion relation of the magnetized plasma results in a cubic equation:

$$\omega_{2n}^3 - \omega_{2n}^2 \omega_b - \omega_{2n}\left[k^2 c^2 + \omega_p^2(t)\right] + k^2 c^2 \omega_b = 0, \tag{8.1}$$

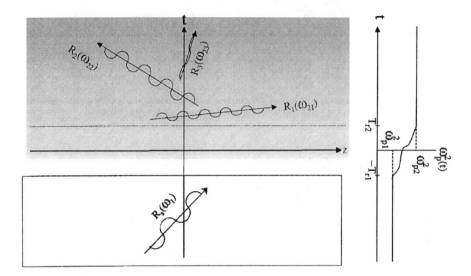

FIGURE 8.1 Geometry of the problem. A right circularly polarized source wave propagating in a magnetized time-varying plasma medium is considered. The direction of wave propagation, which is also the direction of the static magnetic field, is assumed to be along the z-axis.

where

$$k^2 c^2 = \omega_1^2 \left[1 - \frac{\omega_{p1}^2}{\omega_1 \left(\omega_1 - \omega_b \right)} \right]$$
(8.2)

and k is the conserved wave number. The roots of Equation 8.1 are the instantaneous frequencies and are sketched in Figure 8.2.

On the other hand, there is no exact solution for the amplitudes of the waves. A perturbation technique for the computation of the scattering coefficients is the focus of this chapter. A causal Green's function is developed as the basis for the perturbation. The method gives a closed form expression for the scattering coefficients for a general temporal profile for which an exact solution is not available. The extension of the results for the case of a left circularly polarized source wave (L_s wave) is quite straightforward and is not discussed here.

8.2   PERTURBATION FROM THE STEP PROFILE

The governing field equations for the problem, with the assumption of a lossless cold plasma, are

$$\nabla \times \mathbf{E} = -\mu_0 \frac{\partial \mathbf{H}}{\partial t},$$
(8.3)

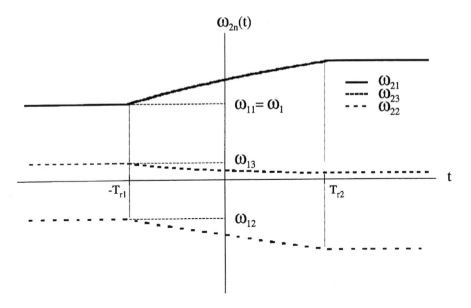

FIGURE 8.2   Instantaneous frequencies.

$$\nabla \times \mathbf{H} = \varepsilon_0 \frac{\partial \mathbf{E}}{\partial t} + \mathbf{J}, \tag{8.4}$$

$$\frac{\partial \mathbf{J}}{\partial t} = \varepsilon_0 \omega_p^2(t)\mathbf{E} - \mathbf{J} \times \mathbf{\omega}_b, \tag{8.5}$$

where $\mathbf{E}$, $\mathbf{H}$, and $\mathbf{J}$ have the usual meaning.[2] For the case of the longitudinal propagation of an R wave, i.e., $\mathbf{H}(z,t) = (\hat{x} - j\hat{y})H(t)e^{-jkz}$, etc.,[2] the magnetic field satisfies

$$\frac{d^3 H(t)}{dt^3} - j\omega_b \frac{d^2 H(t)}{dt^2} + \left[k^2 c^2 + \omega_p^2(t)\right]\frac{dH(t)}{dt} - jk^2 c^2 \omega_b H(t) = 0. \tag{8.6}$$

The solution of Equation 8.4, where $\omega_p^2(t) = \tilde{\omega}_p^2$ is a step function, can be obtained by using the Laplace transform technique:[1]

$$H_0(t) = e^{j\omega_1 t} \qquad\qquad t < 0, \tag{8.7}$$

$$H_0(t) = H_{21}^{(0)} e^{j\omega_{21} t} + H_{22}^{(0)} e^{j\omega_{22} t} + H_{23}^{(0)} e^{j\omega_{23} t} \qquad t > 0, \tag{8.8}$$

where

$$H_{2n}^{(0)} = \frac{\left(\omega_{2n} - \omega_b\right)\left(\omega_{2n} + \omega_1\right) - k^2 c^2 + \omega_1^2 - \omega_{p2}^2}{\displaystyle\prod_{p=1,p\neq n}^{3} \left(\omega_{2n} - \omega_{2p}\right)}, \quad n = 1, 2, 3, \qquad (8.9)$$

ω_1 is the source wave frequency, and ω_{2n} are the roots of

$$\omega_{2n}^3 - \omega_{2n}^2 \omega_b - \omega_{2n}\left(k^2 c^2 + \omega_{p2}^2\right) + k^2 c^2 \omega_b = 0. \qquad (8.10)$$

For a general profile (Figure 8.1) of $\omega_p^2(t)$, a perturbation solution is sought by using the solution for the step profile $\tilde{\omega}_p^2(t)$ as a reference solution. The formulation is developed below:

$$\frac{d^3 H(t)}{dt^3} - j\omega_b \frac{d^2 H(t)}{dt^2} + \left[k^2 c^2 + \omega_p^2(t)\right]\frac{dH(t)}{dt} - jk^2 c^2 \omega_b H(t) = 0 \qquad (8.11)$$

$$\frac{d^3}{dt^3}\left[H_0(t) + H_1(t)\right] - j\omega_b \frac{d^2}{dt^2}\left[H_0(t) + H_1(t)\right]$$

$$+ \left[k^2 c^2 + \tilde{\omega}_p^2(t) + \Delta\omega_p^2(t)\right]\frac{d}{dt}\left[H_0(t) + H_1(t)\right] - jk^2 c^2 \omega_b \left[H_0(t) + H_1(t)\right] = 0. \qquad (8.12)$$

In Equation 8.12

$$\Delta\omega_p^2(t) = \omega_p^2(t) - \tilde{\omega}_p^2(t). \qquad (8.13)$$

The first-order perturbation H_1 can be estimated by dropping the second-order term $-\Delta\omega_p^2(t)d[H_1(t)]/dt$ from Equation 8.12:

$$\frac{d^3 H_1(t)}{dt^3} - j\omega_b \frac{d^2 H_1(t)}{dt^2} + \left[k^2 c^2 + \tilde{\omega}_p^2(t)\right]\frac{dH_1(t)}{dt} - jk^2 c^2 \omega_b H_1(t) = -\Delta\omega_p^2(t)\frac{dH_0(t)}{dt}. \qquad (8.14)$$

By treating the right side of Equation 8.14 as the driving function $F(t)$, the solution of H_1 can be written:

$$H_1(t) = G(t,\tau) * F(t) = -\int_{-\infty}^{t} G(t,\tau)\Delta\omega_p^2(\tau)\frac{dH_0(\tau)}{d\tau}\,d\tau, \qquad (8.15)$$

where $G(t, \tau)$ is the Green's function that satisfies the third-order differential equation:

$$\frac{d^3 G(t,\tau)}{dt^3} - j\omega_b \frac{d^2 G(t,\tau)}{dt^2} + \left[k^2 c^2 + \tilde{\omega}_p^2(t)\right]\frac{dG(t,\tau)}{dt} - jk^2 c^2 \omega_b G(t,\tau) = \delta(t - \tau). \qquad (8.16)$$

The approximate one-iteration solution for the original time-varying differential equation takes the form:

$$H(t) \approx H_0(t) + H_1(t) = H_0(t) - \int_{-\infty}^{t} G(t,\tau) \Delta\omega_p^2(\tau) \frac{dH_0(\tau)}{d\tau} d\tau. \qquad (8.17)$$

More iterations can be made until the desired accuracy is reached. The general N-iterations solution is

$$H(t) = H_0(t) + \sum_{n=1}^{N} H_n(t), \qquad (8.18)$$

where

$$H_n(t) = -\int_{-\infty}^{t} G(t,\tau) \Delta\omega_p^2(\tau) \frac{d}{d\tau} H_{n-1}(\tau) d\tau. \qquad (8.19)$$

8.3 CAUSAL GREEN'S FUNCTION FOR TEMPORALLY UNLIKE MAGNETIZED PLASMA MEDIA

The technique of constructing the Green's function for this case is similar in principle to that used in Reference 7, but is more difficult to construct since the case deals with a third-order system. The frequency of the upshifted forward-propagating wave is different from that of the backward-propagating wave. Also, there is a second forward-propagating wave with a downshifted frequency. The appropriate Green's function in the domain $-\infty < t < 0$ is built up of functions $e^{j\omega_{1m}t}$, where ω_{1m} are the roots of

$$\omega_{1m}^3 - \omega_{1m}^2 \omega_b - \omega_{1m}\left(k^2 c^2 + \omega_{p1}^2\right) + k^2 c^2 \omega_b = 0. \qquad (8.20)$$

One of the roots of the cubic Equation 8.20 is ω_1, say,

$$\omega_{11} = \omega_1. \qquad (8.21)$$

In the region $0 < t < \infty$, the Green's function is built up from the functions $e^{j\omega_{2n}t}$, where ω_{2n} are the roots of Equation 8.10.

From the causality requirement,

$$G(t,\tau) = 0 \quad t < \tau. \qquad (8.22)$$

For $t > \tau$, $G(t, \tau)$ will be determined from the impulse source conditions for a third-order system, i.e.,

$$G(\tau^+, \tau) = G(\tau^-, \tau),\qquad (8.23)$$

$$\frac{\partial G(\tau^+, \tau)}{\partial t} = \frac{\partial G(\tau^-, \tau)}{\partial t},\qquad (8.24)$$

$$\frac{\partial^2 G(\tau^+, \tau)}{\partial t^2} - \frac{\partial^2 G(\tau^-, \tau)}{\partial t^2} = 1.\qquad (8.25)$$

From Equation 8.22, $G(\tau^-, \tau) = 0$, $\partial G(\tau^-, \tau)/\partial t = 0$, and $\partial^2 G(\tau^-, \tau)/\partial t^2 = 0$. Thus, Equations 8.23 through 8.25 reduce to

$$G(\tau^+, \tau) = 0,\qquad (8.26)$$

$$\frac{\partial G(\tau^+, \tau)}{\partial t} = 0,\qquad (8.27)$$

$$\frac{\partial^2 G(\tau^+, \tau)}{\partial t^2} = 1.\qquad (8.28)$$

A geometrical interpretation of the steps involved in constructing the Green's function is given in Figure 8.3, where the horizontal axis is the spatial coordinate z, along which the waves are propagating. The horizontal axis ($t = 0$) is also the temporal boundary between the two magnetized temporally unlike plasma media.

In Figure 8.3a the case $\tau > 0$ is considered. The impulse source is located in the second medium. From Equation 8.16 the Green's function can be written as

$$G(t, \tau) = c_{B1}(\tau)e^{j\omega_{21}t} + c_{B2}(\tau)e^{j\omega_{22}t} + c_{B3}(\tau)e^{j\omega_{23}t}, \qquad t > \tau > 0, \qquad (8.29)$$

and c_{B1}, c_{B2}, and c_{B3} are determined from Equations 8.26 through 8.28:

$$c_{B1}(\tau) = H_{B1}e^{-j\omega_{21}\tau} = \frac{-1}{(\omega_{21} - \omega_{22})(\omega_{21} - \omega_{23})}e^{-j\omega_{21}\tau},\qquad (8.30)$$

$$c_{B2}(\tau) = H_{B2}e^{-j\omega_{22}\tau} = \frac{-1}{(\omega_{22} - \omega_{21})(\omega_{22} - \omega_{23})}e^{-j\omega_{22}\tau}\qquad (8.31)$$

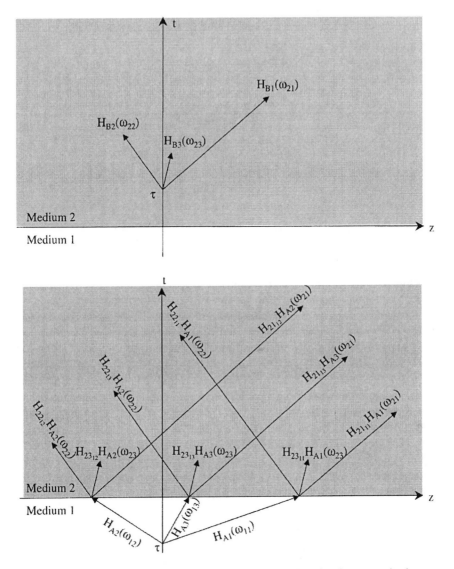

FIGURE 8.3 (a) Geometrical interpretation of the Green's function for magnetized temporally unlike plasma media for the case $\tau > 0$. The frequencies are shown in the parentheses. (b) Geometrical interpretation of the Green's function for magnetized temporally unlike plasma media for the case $\tau < 0$. The frequencies are shown in the parentheses.

$$c_{B3}(\tau) = H_{B3}e^{-j\omega_{23}\tau} = \frac{-1}{(\omega_{23} - \omega_{21})(\omega_{23} - \omega_{22})}e^{-j\omega_{23}\tau}. \qquad (8.32)$$

In Figure 8.3b the case $\tau < 0$ is considered. The impulse source is located in the first medium. Thus, $G(t,\tau)$, in the interval $\tau < t < 0$, can be written as

$$G(t,\tau) = c_{A1}(\tau)e^{j\omega_{11}t} + c_{A2}(\tau)e^{j\omega_{12}t} + c_{A3}(\tau)e^{j\omega_{13}t}, \qquad \tau < t < 0 \qquad (8.33)$$

c_{A1}, c_{A2}, and c_{A3} are once again determined by Equations 8.26 through 8.28:

$$c_{A1}(\tau) = H_{A1}e^{-j\omega_{11}\tau} = \frac{-1}{(\omega_{11} - \omega_{12})(\omega_{11} - \omega_{13})}e^{-j\omega_{11}\tau}, \qquad (8.34)$$

$$c_{A2}(\tau) = H_{A2}e^{-j\omega_{12}\tau} = \frac{-1}{(\omega_{12} - \omega_{11})(\omega_{12} - \omega_{13})}e^{-j\omega_{12}\tau}, \qquad (8.35)$$

$$c_{A3}(\tau) = H_{A3}e^{j\omega_{13}\tau} = \frac{-1}{(\omega_{13} - \omega_{11})(\omega_{13} - \omega_{12})}e^{-j\omega_{13}\tau}. \qquad (8.36)$$

From Figure 8.3b, it is noted that the three waves launched by the impulse source at $t = \tau$ travel along the positive z-axis (two forward-propagating waves) and along the negative z-axis (one backward-propagating wave). They reach the temporal interface at $t = 0$, where the medium undergoes a temporal change. Each of these waves gives rise to three waves at the shifted frequencies ω_{21}, ω_{22}, and ω_{23}. It can be noted that as n varies from 1 to 3 and m varies from 1 to 3, ω_{mn} will have nine values. In actuality, there are only three distinct values for a given wave number k as constrained by the cubic Equation 8.20. The other six values coincide with one of the three distinct values.

The amplitude of the nine waves in the interval $0 < t < \infty$ can be obtained by taking into account the amplitude of each of the incident waves (incident on the temporal interface) and the scattering coefficients. The geometrical interpretation of the amplitude computation of the nine waves is shown in Figure 8.3b. The Green's function for this interval $t > 0$ and $\tau < 0$ is given by

$$G(t,\tau) = \sum_{m=1}^{3}\sum_{n=1}^{3} H_{Am}H_{2n_{1m}}e^{j(\omega_{2n}t - w_{1m}\tau)} \qquad \tau < 0 < t, \qquad (8.37)$$

where

$$H_{2n_{1m}} = \frac{(\omega_{2n} - \omega_{b})(\omega_{2n} + \omega_{1m}) - k^2 c^2 + \omega_{1m} - \omega_{p2}^2}{\displaystyle\prod_{p=1,p\neq n}^{3}(\omega_{2n} - \omega_{2p})}. \qquad (8.38)$$

Explicit expressions for $G(t,\tau)$ valid for various regions of the (t,τ) plane are given in Figure 8.4.

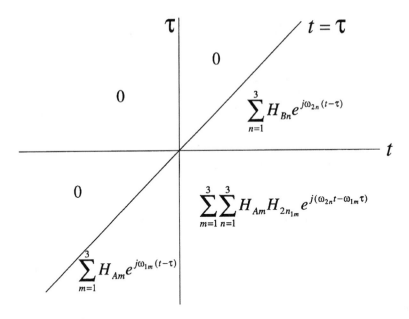

FIGURE 8.4 Causal Green's function for magnetized temporally unlike plasma media.

8.4 SCATTERING COEFFICIENTS FOR A GENERAL PROFILE

The approximate amplitude of the electric field after one iteration ($N = 1$) can be obtained from Equation 8.17:

$$H(t) = H_0(t) - \int_0^{+t} d\tau \sum_{m=1}^{3} H_{Bm} e^{j\omega_{2m}(t-\tau)} \Delta\omega_p^2(\tau) \sum_{n=1}^{3} j\omega_{2n} H_{2n}^{(0)} e^{j\omega_{2n}\tau}$$

$$- \int_{-\infty}^{0} d\tau \sum_{m=1}^{3}\sum_{n=1}^{3} H_{Am} H_{2n_{1m}} e^{j(\omega_{2n}t - \omega_{1m}\tau)} \Delta\omega_p^2(\tau) j\omega_1 e^{j\omega_1\tau}. \tag{8.39}$$

In the asymptotic limit of $t \to \infty$, $H_1(t)$ can be written as

$$H_1(t) = H_{21}^{(1)} e^{j\omega_{21}t} + H_{22}^{(1)} e^{j\omega_{22}t} + H_{23}^{(1)} e^{j\omega_{23}t}, \tag{8.40}$$

where $H_{21}^{(1)}$, $H_{22}^{(1)}$, and $H_{23}^{(1)}$ are the first-order correction terms of the scattering coefficients:

$$H_{21}^{(1)} = -\int_0^{\infty} d\tau H_{B1} e^{-j\omega_{21}\tau} \Delta\omega_p^2(\tau) \sum_{n=1}^{3} j\omega_{2n} H_{2n}^{(0)} e^{j\omega_{2n}\tau}$$

$$- \int_{-\infty}^{0} d\tau \sum_{m=1}^{3} H_{Am} H_{21_{1m}} e^{-j\omega_{1m}\tau} \Delta\omega_p^2(\tau) j\omega_1 e^{j\omega_1\tau}, \tag{8.41}$$

$$H_{22}^{(1)} = -\int_0^\infty d\tau H_{B2} e^{-j\omega_{22}\tau} \Delta\omega^2(\tau) \sum_{n=1}^3 j\omega_{2n} H_{2n}^{(0)} e^{j\omega_{2n}\tau}$$

$$-\int_{-\infty}^0 d\tau \sum_{m=1}^3 H_{Am} H_{22_{1m}} e^{-j\omega_{1m}\tau} \Delta\omega^2(\tau) j\omega_1 e^{j\omega_1\tau}, \tag{8.42}$$

$$H_{23}^{(1)} = -\int_0^\infty d\tau H_{B3} e^{-j\omega_{23}\tau} \Delta\omega_p^2(\tau) \sum_{n=1}^3 j\omega_{2n} H_{2n}^{(0)} e^{j\omega_{2n}\tau}$$

$$-\int_{-\infty}^0 d\tau \sum_{m=1}^3 H_{Am} H_{23_{1m}} e^{-j\omega_{1m}\tau} \Delta\omega_p^2(\tau) j\omega_1 e^{j\omega_1\tau}. \tag{8.43}$$

Higher-order correction terms ($N > 1$) can be obtained by using more iterations.

8.5   SCATTERING COEFFICIENTS FOR A LINEAR PROFILE

As a particular example, Equations 8.41 through 8.43 are used to compute scattering coefficients for a profile of $\omega_p^2(t)$ rising linearly from 0 (at $t = -T_r/2$) to ω_{p2}^2 (at $t = T_r/2$) with a slope of ω_{p2}^2/T_r. Here, T_r is the rise time of the profile. The function $\Delta\omega_p^2(t)$ takes the form of $\omega_{p2}^2(-\frac{1}{2} + t/T_r)$ for $0 < t < T_r/2$ and $\omega_{p2}^2 (\frac{1}{2} + t/T_r)$ for $-T_r/2 < t < 0$. It is zero on the rest of the real line, Figure 8.5.

$$H_{21}^{(1)} = j\omega_{p2}^2 \left[+\frac{\omega_{21} H_{21}^{(0)} H_{B1} T_r}{8} - \omega_{21} H_{22}^{(0)} H_{B1} \frac{e^{j(\omega_{22}-\omega_{21})T_r/2} -1 - j(\omega_{22}-\omega_{21})T_r/2}{T_r(\omega_{22}-\omega_{21})^2} \right.$$

$$-\omega_{21} H_{23}^{(0)} H_{B1} \frac{e^{j(\omega_{23}-\omega_{21})T_r/2} -1 - j(\omega_{23}-\omega_{21})T_r/2}{T_r(\omega_{23}-\omega_{21})^2}$$

$$-\frac{\omega_{11} H_{21_{11}} H_{A1} T_r}{8} + \omega_{12} H_{21_{12}} H_{A2} \frac{e^{-j(\omega_{11}-\omega_{12})T_r/2} -1 + j(\omega_{11}-\omega_{12})T_r/2}{T_r(\omega_{11}-\omega_{12})^2}$$

$$\left. +\omega_{13} H_{21_{13}} H_{A3} \frac{e^{-j(\omega_{11}-\omega_{13})T_r/2} -1 + j(\omega_{11}-\omega_{13})T_r/2}{T_r(\omega_{11}-\omega_{13})^2} \right], \tag{8.44}$$

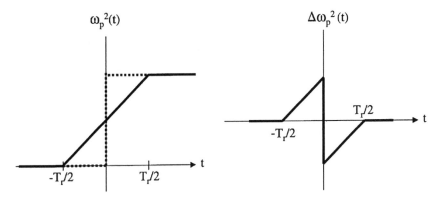

FIGURE 8.5   Linear profile. The broken-line curve is the step profile $\tilde{\omega}_p^2(t)$.

$$H_{22}^{(1)} = j\omega_{p2}^2\left[-\omega_{22}H_{21}^{(0)}H_{B2}\frac{e^{j(\omega_{21}-\omega_{22})T_r/2}-1-j(\omega_{21}-\omega_{22})T_r/2}{T_r(\omega_{21}-\omega_{22})^2}+\frac{\omega_{22}H_{22}^{(0)}H_{B2}T_r}{8}\right.$$

$$-\omega_{22}H_{23}^{(0)}H_{B2}\frac{e^{j(\omega_{23}-\omega_{22})T_r/2}-1-j(\omega_{23}-\omega_{22})T_r/2}{T_r(\omega_{23}-\omega_{22})^2}$$

$$+\omega_{12}H_{22_{12}}H_{A2}\frac{e^{-j(\omega_{11}-\omega_{12})T_r/2}-1+j(\omega_{11}-\omega_{12})T_r/2}{T_r(\omega_{11}-\omega_{12})^2}-\frac{\omega_{11}H_{22_{11}}H_{A1}T_r}{8}$$

$$+\omega_{13}H_{22_{13}}H_{A3}\frac{e^{-j(\omega_{11}-\omega_{13})T_r/2}-1+j(\omega_{11}-\omega_{13})T_r/2}{T_r(\omega_{11}-\omega_{13})^2}\right], \qquad (8.45)$$

$$H_{23}^{(1)} = j\omega_{p2}^2\left[-\omega_{21}H_{21}^{(0)}H_{B3}\frac{e^{j(\omega_{21}-\omega_{23})T_r/2}-1-j(\omega_{21}-\omega_{23})T_r/2}{T_r(\omega_{21}-\omega_{23})^2}\right.$$

$$-\omega_{22}H_{22}^{(0)}H_{B3}\frac{e^{j(\omega_{21}-\omega_{23})T_r/2}-1-j(\omega_{22}-\omega_{23})T_r/2}{T_r(\omega_{22}-\omega_{23})^2}+H_{B3}\frac{\omega_{23}H_{23}^{(0)}T_r}{8}$$

$$+\omega_{11}H_{23_{12}}H_{A2}\frac{e^{-j(\omega_{11}-\omega_{12})T_r/2}-1+j(\omega_{11}-\omega_{12})T_r/2}{T_r(\omega_{11}-\omega_{12})^2}$$

$$+\omega_{11}H_{23_{13}}H_{A3}\frac{e^{-j(\omega_{11}-\omega_{13})T_r/2}-1+j(\omega_{11}-\omega_{13})T_r/2}{T_r(\omega_{11}-\omega_{13})^2}-H_{A1}\frac{\omega_{11}H_{23_{11}}T_r}{8}\right]. \qquad (8.46)$$

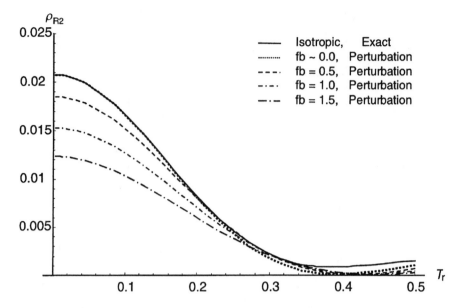

FIGURE 8.6 Power density ρ_{R2} of the backward-propagating wave R_2 vs. rise time for a linear profile.

The solution for other profiles can be obtained similarly. However, to perform the integrations in Equations 8.41 through 8.43 for a general profile, it may be necessary to expand the exponential in Equations 8.41 through 8.43 into a power series and then do the integration. For a given T_r, it is possible to choose the relative positioning of $\omega_p^2(t)$ and $\tilde{\omega}_p^2(t)$, such that

$$\int_{-\infty}^{+\infty} \Delta\omega_p^2(\tau)d\tau = 0. \tag{8.47}$$

Such a choice will improve the accuracy of the scattering coefficients when the exponential terms in Equations 8.41 through 8.43 are approximated by keeping only the dominant terms.

8.6   NUMERICAL RESULTS

Each of the scattering coefficients after one iteration is determined by adding to Equation 8.9 the corresponding first-order correction terms given by Equations 8.43 through 8.46. Figures 8.6, 8.7, and 8.8 depict the power scattering coefficients; the scattering coefficients are first computed from the equations and their magnitude-squares are multiplied by the ratio $|(\omega_1/\omega_{2n})|$ to get the scattering power density. The numerical results for a linear profile are presented in a normalized form by taking $f_1 = \omega_1/2\pi = 1.0$. The other parameters are $f_{p1} = \omega_{p1}/2\pi = 0$ and $f_{p2} = \omega_{p2}/2\pi = 1.2$.

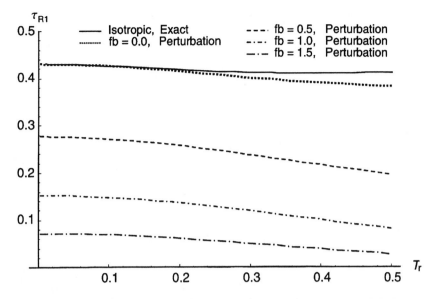

FIGURE 8.7   Power density τ_{R1} of the forward-propagating wave R_1 vs. rise time for a linear profile.

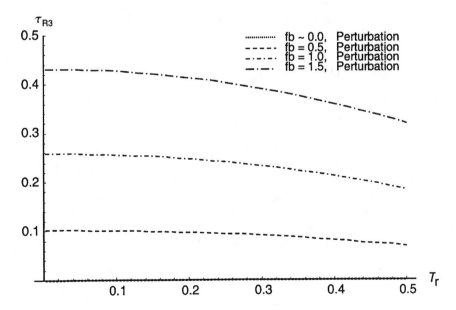

FIGURE 8.8   Power density τ_{R3} of the forward-propagating wave R_3 vs. rise time for a linear profile.

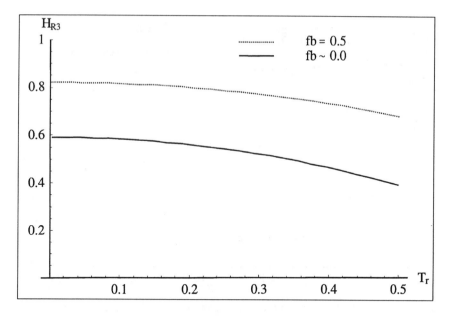

FIGURE 8.9 Magnetic wiggler field vs. rise time.

From these figures, it can be seen that the effect of the rise time is less prominent when the longitudinal magnetic field is strong. From Figure 8.8 it can be noted that the power scattering coefficient of the R_3 wave increases with the strength of the static magnetic field.

8.7 WIGGLER MAGNETIC FIELD

Note that in the $f_b = 0$ case, the perturbation solutions for the R_1 and R_2 waves track the exact isotropic analytical solutions up to $T_r = 0.25$,[7] as expected. The solution for the R_3 wave has a zero power as shown in Figure 8.8. From the sudden-switching analysis,[1] $T_r = 0$, it is known that the electric field for the third wave $E_{23}^{(0)}$ is zero when $f_b = 0$. All the subsequent perturbation correction terms for the electric field are also zero. The magnetic field, however, is not zero,[5] and the wave reduces to a magnetic wiggler field. The perturbation solution for the wiggler field can be calculated from Equation 8.46 by substituting $f_b = 0$. By adding this value to that of $H_{23}^{(0)}$ from Equation 8.9, we get the magnetic wiggler field H_{23} after one iteration is obtained. Thus, the magnetic wiggler field H_w is given by

$$H_w = H_{23}^{(0)} + H_{23}^{(1)}, \qquad f_b = 0. \tag{8.48}$$

Figure 8.9 shows the variation of H_{R3} as a function of T_r. The wiggler magnetic field H_w is obtained by choosing a very small value for f_b and is shown as the solid line. The broken-line curve shows H_{R3} for $f_b = 0.5$. This is the magnetic field of the second forward-propagation wave with a downshifted frequency.

8.8   *E*-FORMULATION

The problem may also be solved using an *E*-formulation. The time-varying differential equation for this formulation is given by Equation 7.9. The results obtained by such a formulation agree with the results shown in Section 8.7.

8.9   SUMMARY

Wave propagation in a transient magnetoplasma with a rise time comparable with the source wave period is a challenging problem. The underlining difficulty is in finding the solution to the third-order governing differential equation with time-varying coefficients. It has no analytical solution even for a linear profile. The development of the Green's function and perturbation series provides a unique approach in providing the analytical insights.

The method has the inherent advantage of using the intrinsic modes of wave propagation, the forward-propagating and the backward-propagating waves, and thus is superior to the numerical solution of the initial value problem. By using this new approach, the variation of the scattering coefficients of the frequency-shifted waves due to a finite rise time of the transient magnetized plasma can be computed individually. The linear profile, which has no analytical solution, illustrates the superiority of the approach.

When ω_b is negligible, the magnetoplasma degenerates into an isotropic plasma. Correspondingly, the magnetoplasma perturbation solution reduces to the isotropic perturbation solution that, in the linear profile case, matches the analytical solution closely. The finite rise time effect on the magnetic wiggler field can also be computed.

The causal Green's function developed here, being fundamental to the interaction of an electromagnetic wave with a magnetized time-varying plasma, can be used in other analyses besides the computation of scattering coefficients.

REFERENCES

1. Kalluri, D. K., Effect of switching a magnetoplasma medium on a traveling wave: longitudinal propagation, *IEEE Trans. Antennas and Propag.*, AP-37, 1638, 1989.
2. Kalluri, D. K., Goteti, V. R., and Sessler, A. M., WKB Solution for wave propagation in a time-varying magnetoplasma medium: longitudinal propagation, *IEEE Trans. Plasma Sci.*, 21, 70, 1993.
3. Goteti, V. R. and Kalluri, D. K., Wave propagation in a switched magnetoplasma medium: transverse propagation, *Radio Sci.*, 25, 61, 1990.
4. Dimitrijevic, M. M. and Stanic, B. V., EMW transformation in suddenly created two-component magnetized plasma, *IEEE Trans. Plasma Sci.*, 23, 422, 1995.
5. Kalluri, D. K., Frequency upshifting with power intensification of a whistler wave by a collapsing plasma medium, *J. Appl. Phys.*, 79, 3895, 1996.
6. Lai, C. H., Katsouleas, T. C., Mori, W. B., and Whittum, D., Frequency upshifting by an ionization front in a magnetized plasma, *IEEE Trans. Plasma Sci.*, 21, 45, 1993.
7. Huang, T. T., Lee, J. H., Kalluri, D. K., and Groves, K. M., Wave propagation in a transient plasma: development of a Green's Function, *IEEE Trans. Plasma Sci.*, 1, 19, 1998.

8. Huang, T. T. and Kalluri, D. K., Effect of finite rise time on the strength of the wiggler magnetic field generated in a switched plasma medium, paper to be presented in 1998 IEEE International Conference on Plasma Science.

9. Kalluri, D. K. and Huang, T. T., Longitudinal propagation in a magnetized time-varying plasma: development of Green's Function, *IEEE Trans. Plasma Sci.* in press.

10. Huang, T. T., Wave Propagation in a Time-Varying Magnetoplasma: Development of a Green's Function, Ph.D. Thesis, University of Massachusetts Lowell, 1997.

9 Adiabatic Analysis of the Modified Source Wave in a Transient Magnetoplasma

9.1 INTRODUCTION

Chapter 7 considered a step profile for the electron density, and Chapter 8 improved the model by allowing a small rise time T_r. This chapter looks at the other limit of approximation of a large rise time. It is assumed that the parameters of the magnetoplasma medium change slowly enough that an *adiabatic analysis* can be made. The objective is to determine the modifications of the source wave caused by the time-varying parameters. Such an analysis for a source R wave will be illustrated in the next section. Similar analysis can be done for the L wave by associating the appropriate algebraic sign with ω_b.

9.2 ADIABATIC ANALYSIS FOR THE R WAVE

The starting point for the adiabatic analysis of the R wave in a time-varying space-invariant magnetoplasma medium (Equation 7.8) is repeated here for convenience.

$$\frac{d^3 H}{dt^3} - j\omega_b(t)\frac{d^2 H}{dt^2} + \left[k^2 c^2 + \omega_p^2(t)\right]\frac{dH}{dt} - j\omega_b(t)k^2 c^2 H = 0. \qquad (9.1)$$

Note that this equation for H does not involve the time derivatives of $\omega_b(t)$ or $\omega_p^2(t)$. The technique[1-4] used in this book in obtaining an adiabatic solution of this equation is as follows. A complex instantaneous frequency function is defined, such that

$$\frac{dH}{dt} = p(t)H(t) = \left[\alpha(t) + j\omega(t)\right]H(t). \qquad (9.2)$$

Here, $\omega(t)$ is the instantaneous frequency. By substituting Equation 9.2 in Equation 9.1 and neglecting α and all derivatives, a zero-order solution can be obtained. The solution gives a cubic in ω, providing the instantaneous frequencies of three waves created by the switching action,

$$\omega^3 - \omega_b(t)\omega^2 - \left[k^2 c^2 + \omega_p^2(t)\right]\omega + k^2 c^2 \omega_b(t) = 0. \qquad (9.3)$$

The cubic has two positive real roots and one negative real root. At $t = 0$, one of the positive roots, say, ω_m, has a value ω_0:

$$\omega_m(0) = \omega_0. \tag{9.4}$$

The mth wave is the modified source wave. An equation for α can now be obtained by substituting Equation 9.2 into Equation 9.1 and equating the real part to zero. In the adiabatic analysis, the derivatives and powers of α are neglected.

$$\alpha = \dot\omega \, \frac{3\omega - \omega_b(t)}{k^2c^2 + \omega_p^2(t) - 3\omega^2 + 2\omega\omega_b(t)}. \tag{9.5}$$

The approximate solution to Equation 9.1 can now be written as

$$H(t) \approx H_m(t) = H_m(t)\exp\left[j\int_0^t \omega_m(\tau)d\tau\right], \tag{9.6}$$

where

$$H_m(t) = H_o \exp\left[\int_0^t \alpha_m(\tau)d\tau\right]. \tag{9.7}$$

The amplitude of the other two modes (waves other than the mth mode) are of the order of the slopes at the origin of $\omega_b(t)$ and $\omega_p^2(t)$ and hence neglected in the adiabatic analysis. The integral in Equation 9.7 can be evaluated numerically, but in the case of only one of the parameters varying with time, it can be evaluated analytically.

In the case of ω_b constant and ω_p^2 varying with time, $\omega_p^2(t)$ can be eliminated from Equation 9.5 by using Equation 9.3 and simplifying:

$$\omega_p^2(t) = \frac{(\omega^2 - k^2c^2)(\omega - \omega_b)}{\omega}, \tag{9.8}$$

$$\alpha = -\dot\omega\left[\frac{3\omega^2 - \omega\omega_b}{2\omega^3 - \omega^2\omega_b - k^2c^2\omega_b}\right], \quad \omega_b = \text{constant}. \tag{9.9}$$

Since the numerator in the bracketed fraction is one half of the derivative of the denominator with respect to ω in the bracketed fraction, Equation 9.9 can be written as

$$\alpha dt = -\frac{1}{2}\frac{dr}{r}, \tag{9.10}$$

where

$$r = 2\omega^3 - \omega^2\omega_b - k^2c^2\omega_b, \omega_b = \text{constant}. \tag{9.11}$$

Equation 9.7 reduces to

$$\frac{H_m(t)}{H_0} = \left[\frac{2\omega_0^3 - \omega_b\omega_0^2 - k^2c^2\omega_b}{2\omega_m^3(t) - \omega_b\omega_m^2(t) - k^2c^2\omega_b} \right]^{1/2}, \quad \omega_b = \text{constant}. \tag{9.12}$$

In the case of ω_p^2 constant and ω_b varying with time, $\omega_b(t)$ can be eliminated from Equation 9.5 by using Equation 9.3 and simplifying

$$\omega_b(t) = \omega\left(1 - \frac{\omega_p^2}{\omega^2 - k^2c^2} \right), \tag{9.13}$$

$$\alpha = -\dot{\omega}\, \frac{\omega\left(2\omega^2 - 2k^2c^2 + \omega_p^2\right)}{\left(\omega^2 - k^2c^2\right)^2 + \omega_p^2\omega^2 + k^2c^2\omega_p^2}. \tag{9.14}$$

As in the previous case, the equation can be integrated analytically and written:

$$\int_0^t \alpha(\tau)d\tau = \ln\left(\frac{s[\omega(0)]}{s[\omega(t)]} \right)^{1/2}, \tag{9.15}$$

where $r(\omega)$ is the denominator in the fraction on the right side of Equation 9.14. From Equation 9.7

$$\frac{H_m(t)}{H_0} = \left[\frac{\left(\omega_0^2 - k^2c^2\right)^2 + \omega_p^2\omega_0^2 + k^2c^2\omega_p^2}{\left(\omega_m^2(t) - k^2c^2\right)^2 + \omega_p^2\omega_m^2(t) + k^2c^2\omega_p^2} \right]^{1/2}. \tag{9.16}$$

The electric field $E_m(t)$ is easily obtained from $H_m(t)$ by using the wave impedance concept:

$$E_m(t) = \eta_{pm}H_m(t), \tag{9.17}$$

$$\eta_{pm} = \frac{\eta_0}{n_m} = \frac{\eta_0\omega_m}{kc}. \tag{9.18}$$

In the above, n_m is the refractive index of the medium when the frequency of the signal in the plasma is ω_m.

9.3 MODIFICATION OF THE SOURCE WAVE BY A SLOWLY CREATED PLASMA

Assume that an R wave of frequency ω_0 is propagating in free space. At $t = 0$, an unbounded homogeneous slowly varying transient plasma with an exponential profile is created:

$$\omega_p^2(t) = \omega_{p0}^2[1 - \exp(-bt)]. \tag{9.19}$$

The fields are given by Equation 9.12 with $kc = \omega_0$. Figure 9.1 shows the variation of the normalized instantaneous frequency, the normalized electric, and the magnetic fields as a function of time. The parameters are $b = 0.1$ and $\omega_0 = 1.0$, and ω_{p0} is chosen as 0.4. A detailed derivation of the equations based on a slightly different formulation is given in Reference 1.

9.4 MODIFICATION OF THE WHISTLER WAVE BY A COLLAPSING PLASMA MEDIUM

Section 6.3 mentioned the whistler mode of propagation of the R wave in the frequency band $0 < \omega < \omega_b$. In this band the refractive index is greater than 1. This section looks at the adiabatic transformation of the whistler wave by a collapsing plasma medium. If n_0 is the refractive index and ω_0 is the frequency of the whistler wave before the collapse begins, the wave magnetic field $H_m(t)$ during the collapse is given by Equation 9.12, where $k^2 c^2 = \omega_0^2 n_0^2$ and the electric field is obtained from Equations 9.17 and 9.18. By denoting the first wave as the modified source wave, i.e., $m = 1$ and $\omega_m(0) = \omega_1(0) = \omega_0$

$$\frac{H_1(t)}{H_0} = \left[\frac{2\omega_0^3 - \omega_b \omega_0^2 - n_0^2 \omega_0^2 \omega_b}{2\omega_1^3(t) - \omega_b \omega_1^2(t) - n_0^2 \omega_0^2 \omega_b} \right]^{1/2}, \tag{9.20}$$

$$\frac{E_1(t)}{E_0} = \frac{\omega_1(t)}{\omega_0} \frac{H_1(t)}{H_0}, \quad \omega_b = \text{constant}. \tag{9.21}$$

It has been mentioned before, see Section 6.3, that the refractive index could be quite large when $\omega_0 \ll \omega_b$. In such a case,

$$n_0 \approx n_w = \frac{\omega_p}{\sqrt{\omega_0 \omega_b}} \gg 1, \tag{9.22}$$

$$\omega_1(t \to \infty) = n_w \omega_0, \tag{9.23}$$

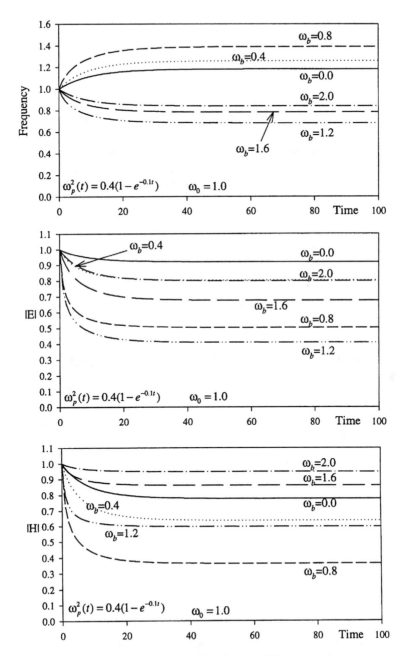

FIGURE 9.1 Frequency and fields of the modified source *R* wave.

$$\frac{E_1(t \to \infty)}{E_0} \approx \frac{n_w}{\sqrt{2}},$$
(9.24)

$$\frac{H_1(t \to \infty)}{H_0} \approx \frac{1}{\sqrt{2}},$$
(9.25)

$$\frac{S_1(t \to \infty)}{S_0} \approx \frac{n_w}{2}.$$
(9.26)

The above equations show that the whistler wave can have a substantial frequency upshift with power intensification of the signal. This aspect is thoroughly examined in Reference 3 for a sudden collapse as well as for a slow decay of the electron density of the plasma medium. Reference 3 is given in this book as Appendix E.

Before leaving this topic, a comment is in order to explain the extra factor

$$\text{Factor} = \frac{\omega_b - \omega_1(t)}{\omega_b - \omega_0}$$
(9.27)

in Equation 14a (see Appendix E) for $E_1(t)/E_0$ as compared with Equation 9.21 of this section. In Appendix E, an alternative physical model for the collapsing plasma is assumed. However, the final result for the whistler mode is still given by Equation 9.22 through 9.26.

9.5   ALTERNATE MODEL FOR A COLLAPSING PLASMA

Equation 9.1 is obtained from a formulation where E, H, J are chosen as the state variables. The third state variable is chosen as the current density field J rather than the velocity field v. Section 1.2 explains this model on the basis of the physical process involved in creating the plasma. The electron density $N(t)$ increases because of the new electrons born at different times. The newly born electrons start with zero velocity and are subsequently accelerated by the fields. Thus, all the electrons do not have the same velocity at a given time during the creation of the plasma. Therefore,

$$J(t) \neq -qN(t)v(t),$$
(9.28)

but

$$\Delta J(t) = -q\Delta N_i v_i(t).$$
(9.29)

In the above, ΔN_i is the electron density added at t_i and $v_i(t)$ is the velocity at t of these ΔN_i electrons born at t_i. $J(t)$ is given by the integral of Equation 9.29 and not by Equation 9.28. $J(t)$ obtained from Equation 9.29 is a smooth function of time and has no discontinuities. It fits the requirements of a state variable. The initial conditions, in this formulation, are the continuity of E, H, and J at $t = 0$. This is a good model for a building-up plasma. The following questions regarding the modeling of a transient plasma may be asked. Is this the only model or are there other models? Is this a good model for the collapsing plasma? A lot depends on the physical processes responsible for the temporal change in the electron density.[5]

Consider the following model for the collapse of the plasma. The decrease in electron density takes place by a process of sudden removal of $\Delta N(t)$ electrons; the velocities of all of the remaining electrons are unaffected by this capture and have the same instantaneous value $v(t)$. The initial conditions in this model are the continuity of E, H, and v; these field variables can be used as the state variables. The state variable equations are

$$\mu_0 \frac{dH}{dt} = jkE, \tag{9.30}$$

$$\varepsilon_0 \frac{dE}{dt} = jkH + \varepsilon_0 \omega_p^2(t)u, \tag{9.31}$$

$$\frac{du}{dt} = -E + j\omega_b(t)u, \tag{9.32}$$

where

$$u = e\frac{v}{m}, \tag{9.33}$$

and the simplest higher-order differential equation, when ω_b is a constant, is in the variable u and is given by

$$\frac{d^3u}{dt^3} - j\omega_b \frac{d^2u}{dt^2} + \left[k^2c^2 + \omega_p^2(t)\right]\frac{du}{dt} + \left[g(t) - j\omega_b k^2c^2\right]u = 0, \quad \omega_b = \text{constant}, \tag{9.34}$$

where

$$g(t) = \frac{d\omega_p^2(t)}{dt}. \tag{9.35}$$

Equation 9.34 is slightly different from Equation 9.1. The last term on the left side of Equation 9.34 has an extra parameter, $g(t)$. Consequently, whereas the equation for ω remains the same as Equation 9.3, the equation for α is a modified version of Equation 9.5 and is given by Equation 9.36:

$$\alpha = \frac{-g(t) + \dot{\omega}(3\omega - \omega_b)}{k^2 c^2 + \omega_p^2(t) - 3\omega^2 + 2\omega\omega_b}, \quad \omega_b = \text{constant}. \tag{9.36}$$

By substituting Equation 9.8 for $\omega_p^2(t)$ in Equation 9.36,

$$\alpha = \frac{g(t)\omega - \dot{\omega}(3\omega^2 - \omega_b\omega)}{2\omega^3 - \omega^2\omega_b - \omega_b k^2 c^2}, \quad \omega_b = \text{constant}. \tag{9.37}$$

An expression for $g(t)$ can be obtained by differentiating Equation 9.8:

$$g(t) = \frac{d\omega_p^2(t)}{dt} = \frac{\dot{\omega}}{\omega}\left[2\omega^3 - \omega^2\omega_b - k^2 c^2\omega_b\right], \quad \omega_b - \text{constant}. \tag{9.38}$$

By substituting Equation 9.38 in Equation 9.37,

$$\alpha = \frac{\dot{\omega}}{\omega} - \dot{\omega}\frac{\omega^2 - \omega\omega_b}{2\omega^3 - \omega^2\omega_b - k^2 c^2\omega_b}, \quad \omega_b = \text{constant}. \tag{9.39}$$

The velocity field of the modified source wave is obtained, by analytical integration, as before:

$$\frac{u_1(t)}{u_0} = \exp\left[\int_0^t \alpha(\tau)d\tau\right] = \frac{\omega_1(t)}{\omega_0}\left[\frac{2\omega_0^3 - \omega_0^2\omega_b - n_0^2\omega_0^2\omega_b}{2\omega_1^3 - \omega_1^2\omega_b - n_0^2\omega_0^2\omega_b}\right]^{1/2}. \tag{9.40}$$

The electric field (Equation 9.43) is now obtained by noting from Equation 9.32

$$E_0 = j(\omega_b - \omega_0)u_0, \tag{9.41}$$

$$E_1 = j(\omega_b - \omega_1)u_1, \tag{9.42}$$

$$\frac{E_1(t)}{E_0} = \left[\frac{\omega_b - \omega_1(t)}{\omega_b - \omega_0}\right]\left[\frac{\omega_1(t)}{\omega_0}\right]\left[\frac{2\omega_0^3 - \omega_0^2\omega_b - n_0^2\omega_0^2\omega_b}{2\omega_1^3(t) - \omega_1^2(t)\omega_b - n_0^2\omega_0^2\omega_b}\right]^{1/2}, \quad \omega_b = \text{constant}. \tag{9.43}$$

When $\omega_0 \ll \omega_b \sim \omega_p$, $n_0 \approx n_w$ is large and the first bracketed term on the right side of Equation 9.43 can be approximated by

$$\left[\frac{\omega_b - \omega_1(t)}{\omega_b - \omega_0}\right] \approx \left[1 + \frac{\omega_0}{\omega_b} - \frac{\omega_{p0}}{\omega_b}\sqrt{\frac{\omega_0}{\omega_b}}\right] \approx 1. \tag{9.44}$$

9.6   MODIFICATION OF THE WHISTLER WAVE BY A COLLAPSING MAGNETIC FIELD

Equation 9.16 can be directly applied to this case. The third mode is denoted as the modified source wave, i.e., $\omega_3(0) = \omega_0$ and $kc = n_0\omega_0$:

$$\frac{H_3(t)}{H_0} = \sqrt{\frac{\left(\omega_0^2 - n_0^2\omega_0^2\right)^2 + \omega_p^2\omega_0^2 + n_0^2\omega_0^2\omega_p^2}{\left(\omega_3^2(t) - n_0^2\omega_0^2\right)^2 + \omega_p^2\omega_3^2(t) + n_0^2\omega_0^2\omega_p^2}}. \tag{9.45}$$

In this case it is immaterial whether J or v is chosen as the state variable since the continuity of v ensures the continuity of J, the electron density N being a constant.

For the case when $\omega_0 \ll \omega_b \sim \omega_p$, $n_0 \approx n_w$ is large, and it can be shown:

$$H_2(\infty) \approx H_0, \tag{9.46}$$

$$E_3(\infty) \approx 0, \tag{9.47}$$

$$\omega_3(\infty) \approx 0. \tag{9.48}$$

The whistler wave is converted into a wiggler magnetic field. This aspect is discussed at length in Reference 4 and is given in this book as Appendix F.

9.7   ADIABATIC ANALYSIS FOR THE X WAVE

This chapter discussed the adiabatic analysis of the R wave in a transient magneto-plasma. A similar analysis can be performed for the X wave[2] and is given in this book as Appendix H.

REFERENCES

1. Kalluri, D. K., Goteti, V. R., and Sessler, A. M., WKB solution for wave propagation in a time-varying magnetoplasma medium: longitudinal propagation, *IEEE Trans. Plasma Sci.*, 21, 70, 1993.
2. Lee, J. H. and Kalluri, D. K., Modification of an electromagnetic wave by a time-varying switched magnetoplasma medium: transverse propagation, *IEEE Trans. Plasma Sci.*, 26, 1, 1998.
3. Kalluri, D. K., Frequency upshifting with power intensification of a whistler wave by a collapsing plasma medium, *J. Appl. Phys.*, 79, 3895, 1996.
4. Kalluri, D. K., Conversion of a whistler wave into a controllable helical wiggler magnetic field, *J. Appl. Phys.*, 79, 6770, 1996.
5. Stepanov, N. S., Dielectric constant of unsteady plasma, *Sov. Radiophys. Quant. Electr.*, 19, 683, 1976.

10 Miscellaneous Topics

10.1 INTRODUCTION

The first nine chapters have dealt with topics in an orderly fashion progressing step by step from one kind of complexity of the medium to another. In the process the reader, it is hoped, has developed a feel for the theory of *Frequency Shifting by a Transient Magnetoplasma Medium.*

This chapter will consider miscellaneous topics connected with the general theme of the book. They include references to the contributions of some of the research groups with which the author is familiar.

10.2 PROOF-OF-PRINCIPLE EXPERIMENTS

We considered earlier two distinct cases:

Case 1. Time-invariant space-varying medium: This problem was solved by imposing the requirement that ω is conserved (Chapter 2). In the ω–k diagram, ω = constant is a horizontal line.

Case 2. Space-invariant time-varying medium: The conserved quantity now is k and k = constant is a vertical line in the ω–k diagram.

Experimental realization of Case 2 is not easy. A large region of space had to be ionized uniformly at a given time. Joshi et al.,[1] Kuo,[2] Kuo and Sen,[3] and Rader and Alexeff[4] developed ingenious experimental techniques to achieve these conditions and demonstrated the principle of frequency shifting using an isotropic plasma. One of the earliest pieces of experimental evidence of frequency shifting quoted in the literature is the seminal paper by Yablonovitch.[5]

10.3 MOVING IONIZATION FRONT

The Doppler frequency shift associated with reflection of an electromagnetic signal by a moving object (medium) is familiar to all. In this case, the medium along with the boundary is moving. An alternative is to create a moving boundary between two media that themselves do not move. Imagine an electromagnetic signal propagating through an unionized gas. At $t = 0$, a powerful laser pulse is injected into the gas. As the pulse travels with the speed of light through the gas, say, from right to left, it ionizes the gas and creates a plasma. At any given time, there exists a boundary between the unionized gas, which is like free space, and the plasma. The boundary itself is moving with the speed of light. Such a medium is called an ionization front.

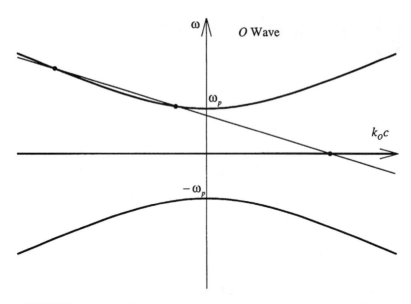

FIGURE 10.1 ω–k diagram of the O wave for an ionization front problem.

Figure 1.2 shows the geometry of the ionization front where the velocity of the front is denoted as v_F. In space-time the front is described by the equation

$$z = -v_F t. \tag{10.1}$$

Let ω_0 and k_0 be the frequency and the wave number in free space to the left of the front and ω and k be the corresponding entities in the plasma to the right of the front. From the phase invariance principle,

$$\left(\omega_0 t - k_0 z\right)\big|z = -v_F t = \left(\omega t - kz\right)\big|z = -v_F t, \tag{10.2}$$

$$\omega = -v_F k + \omega_0(1+\beta), \tag{10.3}$$

$$\beta = v_F/c. \tag{10.4}$$

The equation of the front in the ω–k diagram is given by Equation 10.3, which is a straight line with the slope $-v_F$. The intersection points of this straight line with the ω–k diagram for the plasma determine the frequencies of the new waves created by the ionization front. Figures 10.1 to 10.4 show these intersection points for the O, L, R, and X waves. The associated scattering coefficients are obtained by imposing the requirements on the electric and magnetic fields at a moving boundary. Lai et al.[6] investigated this problem theoretically in detail. A considerable amount of theoretical and experimental work on moving ionization fronts has been done recently.[7-16]

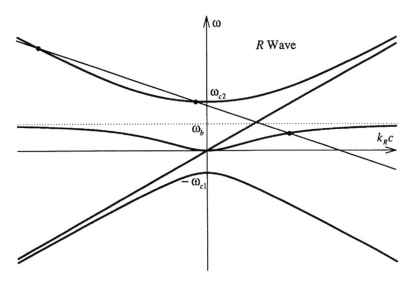

FIGURE 10.2 ω–k diagram of the R wave for an ionization front problem.

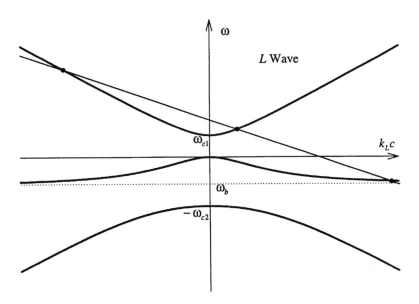

FIGURE 10.3 ω–k diagram of the L wave for an ionization front problem.

Mathematically, the sudden creation problem is a particular case of the moving front problem with $v_F = \infty$; flash ionization is another name for the sudden creation. On the other hand, $v_F = 0$ simulates the stationary boundary problem.

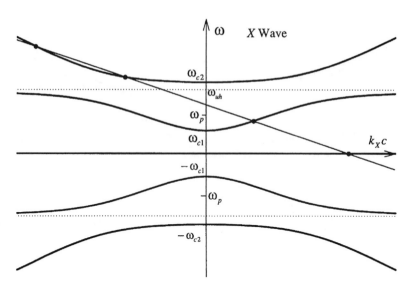

FIGURE 10.4 ω–k diagram of the X wave for an ionization front problem.

10.4 FD-TD METHOD

The previous chapters chose suitable models and made approximations that permitted exact or approximate analytical solutions. These solutions are useful in providing a qualitative picture of the effects of each type of complexity in the properties of the medium. Numerical methods are useful in studying the effects of the simultaneous presence of several complexities. The finite difference time domain technique[17-20] seems to be well suited for handling the problem of transformation of an electromagnetic wave by a time-varying and space-varying magnetoplasma medium.

10.5 LORENTZ MEDIUM

In this book the word dielectric is used to denote a medium whose dielectric constant is a real scalar constant ε_r greater than 1. Following the notation used by Stanic,[21] a dispersive dielectric is denoted by the name Lorentz medium. An electron in a Lorentz medium experiences an additional quasi-elastic restitution force[21] proportional to the displacement, and the equation of motion of the electron Equation 1.3 in such a medium is modified as

$$m\frac{d\mathbf{v}_L}{dt} = -e\mathbf{E} - \alpha \int_0^t \mathbf{v}_L d\tau. \qquad (10.5)$$

The dielectric constant of such a medium denoted by ε_L is given by

$$\varepsilon_L = 1 - \frac{\omega_p^2}{\omega^2 + \omega_L^2}, \qquad (10.6)$$

where

$$\omega_L = \left(\frac{\alpha}{m}\right)^{1/2}. \tag{10.7}$$

For $\alpha = 0$, $\omega_L = 0$, the electron is free and the Lorentz medium becomes a cold plasma.

Stanic[21] and Stanic and Draganic[22] extended the analysis of Chapters 3 and 7 to a Lorentz medium and gave results for the frequencies, fields, and the power densities of various modes. The techniques developed in this book for the time-varying media can be applied to more complex media with multiple Lorentz resonances.[17,18]

10.6   MODE CONVERSION OF THE X WAVE

It is known from the magnetoionic theory that the X wave has a resonance at the upper hybrid frequency

$$\omega_{UH} = \sqrt{\omega_p^2 + \omega_b^2}. \tag{10.8}$$

However, using warm-electron gas equations, it can be shown that in place of the resonance at ω_{UH}, there is a transition from a basically transverse wave to a basically longitudinal wave at this frequency. The ω–k diagram for such a medium can be obtained from the known expression[23] for the refractive index of the X wave:

$$n_X^2 = \frac{\left(\omega^2 - \omega_p^2\right)\left(\omega^2 - \omega_{UH}^2 - \delta^2 k^2 c^2\right) - \omega_p^2 \omega_b^2}{\left(\omega^2 - \omega_{UH}^2 - \delta^2 k^2 c^2\right)}, \tag{10.9}$$

and the relation

$$k^2 c^2 = \omega^2 n_X^2. \tag{10.10}$$

In the above,

$$\delta^2 = \frac{c_e^2}{c^2} = \frac{\gamma k_B T_e}{m}\frac{1}{c^2}, \tag{10.11}$$

where γ, k_B, T_e, and m have the usual meanings[23] and c_e is the electron thermal speed. From Equation 10.9, it is obvious that the upper hybrid resonance is eliminated by the thermal term.

Figure 10.5 shows the ω–k diagram for $\delta = 0.01$, $\omega_0 = 1$, $\omega_p = 0.2$, and $\omega_b = 0.5$. In Figures 10.5 through 10.8, the frequencies are normalized with respect to the source frequency ω_0. The solid line is the diagram for the cold plasma approximation. Figures 10.6 through 10.8 show the ω–k diagram for other values of the parameters

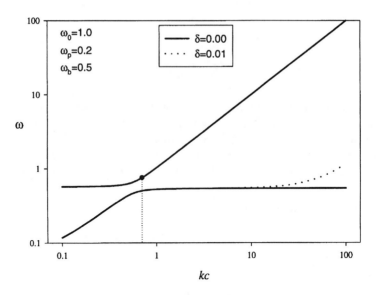

FIGURE 10.5   Mode conversion problem. $\omega-k$ diagram of the X wave for $\omega_p = 0.2$.

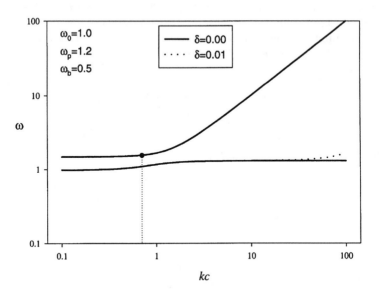

FIGURE 10.6   Mode conversion problem. $\omega-k$ diagram of the X wave for $\omega_p = 1.2$.

ω_p and ω_b. These figures offer the qualitative proof on the possibilities of mode conversion in a time-varying magnetoplasma medium. See Appendix H for a discussion on the possibilities of mode conversion, where the modification of an X wave by a slowly varying magnetoplasma parameter is considered.

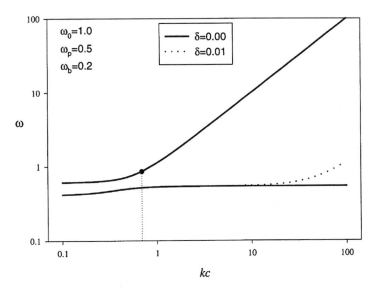

FIGURE 10.7 Mode conversion problem. ω–k diagram of the X wave for $\omega_p = 0.2$.

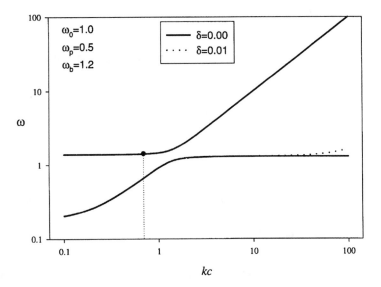

FIGURE 10.8 Mode conversion problem. ω–k diagram of the X wave for $\omega_p = 1.2$.

No work has yet been done on mode conversion in a time-varying magnetoplasma medium. However, some literature exists on the mode conversion in a space-varying magnetoplasma medium.[23] Some literature also exists on such a phenomenon in a magnetoelastic medium.[24,25]

10.7 FREQUENCY-SHIFTING TOPICS OF CURRENT RESEARCH INTEREST

Kuo et al.[26,27] dealt with the following topics:

1. Frequency downshifting due to collisionality of the plasma.[26]
2. Frequency downshifting due to the switching of a periodic medium consisting of plasma layers.[27] The frequencies of the waves generated by the sudden switching of unbounded periodic plasma layers can be obtained by drawing a vertical line $k = $ constant in the $\omega - \beta$ diagram (Figure 2.12).

Mendonca and Silva[28] dealt with the mode coupling theory due to a time-varying and space-varying plasma in a cavity. Goteti and his group[29,30] extended the work of Chapter 5 on switching of the plasma slab to determining the fields in the plasma region in the isotropic case as well as the case of transverse propagation in a magnetized plasma. Nerukh[31] considered an evolutionary approach in transient electrodynamic problems. Muggli et al.[32] recently conducted a proof-of-principle experiment on radiation generation by an ionization front in a gas-filled capacitor array. Yoshii et al.[33] discussed the generation of tunable microwaves from Cerenkov wakes in a magnetized plasma.

10.8 CHIRAL MEDIA: *R* AND *L* WAVES

Chiral materials have molecules of helical structure. They are bi-isotropic materials and have the complex constitutive relations,

$$\mathbf{D} = \varepsilon\mathbf{E} - j\xi_c\mathbf{B} \quad \text{and} \tag{10.12}$$

$$\mathbf{H} = -j\xi_c\mathbf{E} + \frac{1}{\mu}\mathbf{B}, \tag{10.13}$$

where ε and μ are the scalar permittivity and permeability of the media and ξ_c is the chirality parameter, which has the dimensions of admittance. By assuming a one-dimensional harmonic variation of field variables $f(z,t)$,

$$f(z,t) = \exp\left[j\left(\omega t - k_c z\right)\right], \tag{10.14}$$

it can be shown that the R and L waves are the natural modes with the wave number[34]

$$k_c = k_{R,L} = \pm\omega\mu\xi_c + \left[k^2 + \left(\omega\mu\xi_c\right)^2\right]^{1/2}, \tag{10.15}$$

$$k = \omega\sqrt{\mu\varepsilon}, \tag{10.16}$$

where the upper sign is for the R wave and the lower sign is for the L wave. The wave impedance is given by

$$\eta_c = \frac{E_x}{H_y} = \left(\frac{\mu}{\varepsilon + \mu \xi_c^2} \right)^{1/2}. \tag{10.17}$$

Since the R and L waves have different wave numbers, the propagation of a linearly polarized wave is accompanied by the phenomenon of rotation of the plane of polarization. Liquids, such as sugar solutions, and crystals, such as quartz and sodium chloride,[34] exhibit *natural rotation* at optical frequencies due to *optical activity*. Artificial chiral materials that exhibit such rotations at microwave frequencies are commercially available.[35] The natural rotation is similar in principle to the Faraday rotation (magnetic rotation) discussed in Section 6.3. However, the phenomena have the following important difference. In the case of natural rotation, the sign of the angle of rotation is dependent on whether the wave is a forward-propagating wave or a backward-propagating wave. In the magnetic rotation, it depends on the direction of the static magnetic field and does not depend on whether the wave is a forward-propagating wave or a backward-propagating wave. Therefore, the total angle of rotation for a round-trip travel of the wave is zero for the natural rotation. On the other hand, the total angle of rotation for a round-trip travel of the wave for the magnetic rotation is double that of a single-trip travel. An optical isolator is constructed using these rotators.[36]

The theory developed in Section 2.7 can be used to study the wave propagation in an unbounded layered chiral media. Equation 2.127 can be used with the replacement of n by the equivalent refractive index n_c of the chiral medium. Reference 37 used such a technique to study the filter characteristics of periodic chiral layers.

10.9 SOLITONS[Kp38]

This topic deals with a desirable aspect of the EMW transformation by the balancing act of two complexities each of which by itself produces an undesirable effect. The balancing of the wave-breaking feature of a nonlinear medium complexity with the flattening feature of the dispersion leads to a soliton solution.

REFERENCES

1. Joshi, C. J., Clayton, C. E., Marsh, K., Hopkins D. B., Sessler, A., and Whittum, D., Demonstration of the frequency upshifting of microwave radiation by rapid plasma creation, *IEEE Trans. Plasma Sci.*, 18, 814, 1990.
2. Kuo, S. P., Frequency up-conversion of microwave pulse in a rapidly growing plasma, *Phys. Rev. Lett.*, 65, 1000, 1990.
3. Kuo, S. P. and Ren, A., Experimental study of wave propagation through a rapidly created plasma, *IEEE Trans. Plasma Sci.*, 21, 53, 1993.
4. Rader, M. and Alexeff, I., Microwave frequency shifting using photon acceleration, *Int. J. Infrared Millimeter Waves*, 12, 7, 683, 1991.

5. Yablonovitch, E., Spectral broadening in the light transmitted through a rapidly growing plasma, *Phys. Rev. Lett.*, 31, 877, 1973.

6. Lai, C. H., Katsouleas, T. C., Mori, W. B., and Whittum D., Frequency upshifting by an ionization front in a magnetized plasma, *IEEE Trans. Plasma Sci.*, 21, 45, 1993.

7. Lampe, M., Ott, E., Manheimer, W. M., and Kainer, S., Submillimeter-wave production by upshifted reflection from a moving ionization front, *IEEE Trans. Microwave Theor. Tech.*, 25, 556, 1991.

8. Esarey, E., Joyce, G., and Sprangle, P., Frequency up-shifting of laser pulses by copropagating ionization fronts, *Phys. Rev. A*, 44, 3908, 1991.

9. Esarey, E., Ting, A., and Sprangle, P., Frequency shifts induced in laser pulses by plasma waves, *Phys. Rev. A*, 42, 3526, 1990.

10. Lampe, M., Ott, E., and Walker, J. H., Interaction of electromagnetic waves with a moving ionization front, *Phys. Fluids*, 21, 42, 1978.

11. Mori, W. B., Generation of tunable radiation using an underdense ionization front, *Phys. Rev. A*, 44, 5121, 1991.

12. Savage, R. L., Jr., Joshi, C. J., and Mori, W. B., Frequency up-conversion of electromagnetic radiation upon transmission into an ionization front, *Phys. Rev. Lett.*, 68, 946, 1992.

13. Wilks, S. C., Dawson, J. M., and Mori W. B., Frequency up-conversion of electromagnetic radiation with use of an overdense plasma, *Phys. Rev. Lett.*, 63, 337, 1988.

14. Gildenburg, V. B., Pozdnyakova, V. I., and Shereshevski, I. A., Frequency self-upshifting of focused electromagnetic pulse producing gas ionization, *Phys. Lett. A*, 43, 214, 1995.

15. Xiong, C., Yang Z., and Liu, S., Frequency shifts by beam-driven plasma wake waves, *Phys. Lett. A*, 43, 203, 1995.

16. Savage, R. L., Jr., Brogle, R. P., Mori, W. B., and Joshi, C. J., Frequency upshifting and pulse compression via underdense relativistic ionization fronts, *IEEE Trans. Plasma Sci.*, 21, 5, 1993.

17. Taflove, A., *Computational Electrodynamics*, Artech House, Boston, 1995.

18. Kunz, K. S. and Luebbers, R. J., *Finite Difference Time Domain Method for Electromagnetics*, CRC, Boca Raton, FL, 1993.

19. Stanic, B. V., Jelanak, A., and Jelanak, Z., FDTD simulation of dirac impulse propagation in suddenly created plasma, Paper presented at *IEEE Int. Conf. Plasma Sci.*, 1995.

20. Kalluri, D. K. and Lee, J. H., Numerical Simulation of Electromagnetic Wave Transformation in a Dynamic Magnetized Plasma, Final Report, Summer Research Extension Program AFOSR, December 1997.

21. Stanic, B. V., Electromagnetic waves in a suddenly created Lorentz medium and plasma, *J. Appl. Phys.*, 70, 1987, 1991.

22. Stanic, B. V. and Draganic, I. N., Electromagnetic waves in a suddenly created magnetized Lorentz medium (transverse propagation), *IEEE Trans. Antennas Propag.*, AP-44, 1394, 1996.

23. Swanson, D. G., *Plasma Waves*, Academic Press, New York, 1989.

24. Auld, B. A., Collins, J. H., and Japp, H. R., Signal processing in a nonperiodically time-varying magnetoelastic medium, *Proc. IEEE*, 56, 258, 1968.

25. Rezende. S. M., Magnetoelastic and Magnetostatic Waves in a Time-varying Magnetic Fields, Ph.D. Dissertation, M.I.T., Cambridge, MA, 1967.

26. Kuo, S. P., Ren, A., and Schmidt, G., Frequency downshift in a rapidly ionizing media, *Phys. Rev. E*, 49, 3310, 1994.

27. Kuo, S. P. and Faith, J., Interaction of an electromagnetic wave with a rapidly created spatially periodic plasma, *Phys. Rev. E*, 56, 1, 1997.

28. Mendonca, J. T. and Silva, L. O., Mode coupling theory of flash ionization in a cavity, *IEEE Trans. Plasma Sci.*, 24, 147, 1996.

29. Kapoor, S., Wave Propagation in a Bounded Suddenly Created Isotropic Plasma Medium: Transverse Propagation, M.S. Thesis, Tuskegee University, AL, 1992.

30. Dillon, W. J., The Effects of a Sudden Creation of a Magnetoplasma Slab on Wave Propagation: Transverse Propagation, M.S. Thesis, Tuskegee University, AL, 1996.

31. Nerukh, A. G, Evolutionary approach in transient electrodynamic problems, *Radio Sci.*, 30, 481, 1995.

32. Muggli, P., Liou, R., Lai, C. H., and Katsouleas, T. C., Radiation generation by an ionization front in a gas-filled capacitor array, in *Conf. Rec. Abstr., 1997 IEEE Int. Conf. Plasma Sci.*, San Diego, 19–22, May 1997, 157.

33. Yoshii, J., Lai, C. H., and Katsouleas, T. C., Tunable microwaves from Cerenkov wakes in magnetized plasma, in *Conf. Rec. Abstr., IEEE Int. Conf. Plasma Sci.*, San Diego, 1997, 157.

34. Ishimaru, A., *Electromagnetic Wave Propagation, Radiation, and Scattering*, Prentice-Hall, Englewood Cliffs, NJ, 1991.

35. Lindell, I. V., Sihvola, A. H., Tretyakov, S. A., and Viitanen, A. J., *Electromagnetic Waves in Chiral and Bi-Isotropic Media*, Artech House, Boston, 1994.

36. Saleh, B. E. A. and Teich, M. C., *Fundamentals of Photonics*, Wiley, New York, 1991.

37. Kalluri, D. K. and Rao, T. C. K., Filter characteristics of periodic chiral layers, *Pure Appl. Opt.*, 3, 231, 1994.

38. Hirose, A. and Longren, K. E., *Introduction to Wave Phenomena*, John Wiley & Sons, New York, 1985.

ADDITIONAL REFERENCES ADDED IN PROOF

1. Nerukh, A. G., Scherbatko, I. V., and Rybin, O. N., The direct numerical calculation of an integral volterra equation for an electromagnetic signal in a time-varying dissipative medium, *Journal of Electromagnetic Waves and Applications*, 12, 163, 1998.

2. Nerukh, A. G., Scherbatko, I. V., Tyhnenko, A. G., Nikita, K. S. and Uzunoglu, N. K., Scattering of radiation by an object located near the interface of a nonstationary medium, 1, 79, 1997.

3. Nerukh, A. G. and Khizhnyak, N. A., Enhanced reflection of an electromagnetic wave from a plasma cluster moving in a waveguide, *Microwave and Optical Technology Letters*, 17, 267, 1998.

4. Nerukh, A. G., Scherbatko, I. V., and Nerukh, D. A., Using evolutionary recursion to solve an electromagnetic problem with time-varying parameters, *Microwave and Optical Technology Letters*, 14, 31, 1997.

5. Nerukh, A. G., Splitting of an electromagnetic pulse during a jump in the conductivity of a bounded medium (English transl.), *Soviet Physics-Technical Physics*, 37, 543, 1992.

6. Nerukh, A. G., On the transformation of electromagnetic waves by a nonuniformly moving boundary between two media (English transl.), *Soviet Physics-Technical Physics*, 34, 281, 1989.

7. Nerukh, A. G., Electromagnetic waves in dielectric layers with time-dependent parameters (English transl.), *Soviet Physics-Technical Physics*, 32, 1258, 1987.

Appendix A
Constitutive Relation for a
Time-Varying Plasma Medium

The electron density profile function in Figure A.1a $N(\mathbf{r},t)$ can be written as the sum of steady density $N_0(\mathbf{r})$ and the time-varying density $N_1(\mathbf{r},t)$.

$$N(\mathbf{r},t) = N_0(\mathbf{r}) + N_1(\mathbf{r},t). \tag{A.1}$$

Let $\mathbf{J}_1(\mathbf{r},t)$ be the current density due to $N_1(\mathbf{r},t)$. Note that $N_1(\mathbf{r},0) = 0$, as shown in Figure A.1b. From Equations 1.4 through 1.6,

$$\Delta\mathbf{J}_1(\mathbf{r},t) = \frac{q^2}{m}\left\{\frac{\partial N_1}{\partial t}\Big|_{t=t_i}\right\}\Delta t_i \int_{t_i}^{t}\mathbf{E}(\mathbf{r},\alpha)d\alpha, \tag{A.2}$$

$$\mathbf{J}_1(\mathbf{r},t) = \frac{q^2}{m}\int_0^t\frac{\partial N_1(\mathbf{r},\tau)}{\partial\tau}d\tau\int_\tau^t\mathbf{E}(\mathbf{r},\alpha)d\alpha. \tag{A.3}$$

Let

$$\mathbf{f}(\mathbf{r},t) = \int_0^t\frac{\partial N_1(\mathbf{r},\tau)}{\partial\tau}d\tau\int_\tau^t\mathbf{E}(\mathbf{r},\alpha)d\alpha \tag{A.4}$$

and

$$\mathbf{p} = \int\mathbf{E}(\mathbf{r},\alpha)d\alpha, \tag{A.5}$$

$$\int_\tau^t\mathbf{E}(\mathbf{r},\alpha)d\alpha = \mathbf{p}(\mathbf{r},t) - \mathbf{p}(\mathbf{r},\tau), \tag{A.6}$$

$$\mathbf{f}(\mathbf{r},t) = \mathbf{p}(\mathbf{r},t)\int_0^t\frac{\partial N_1(\mathbf{r},\tau)}{\partial\tau}d\tau - \int_0^t\mathbf{p}(\mathbf{r},\tau)\frac{\partial N_1(\mathbf{r},\tau)}{\partial\tau}d\tau. \tag{A.7}$$

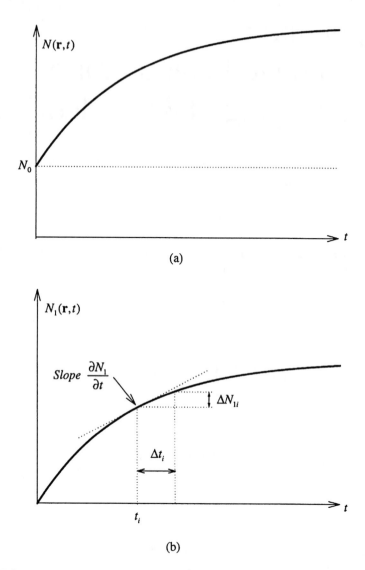

FIGURE A.1 Time-varying electron density profile. N_0 is the initial value and N_1 is the time-varying component.

Integrating the second integral on the right side of Equation A.7 by parts gives

$$\mathbf{f}(\mathbf{r},t) = \mathbf{p}(\mathbf{r},t)\left[N_1(\mathbf{r},t) - N_1(\mathbf{r},0)\right] - \mathbf{p}(\mathbf{r},\tau)N_1(\mathbf{r},\tau)\big|_0^t + \int_0^t N_1(\mathbf{r},\tau)\frac{\partial \mathbf{p}(\mathbf{r},\tau)}{\partial \tau}d\tau. \quad (A.8)$$

Since $N_1(\mathbf{r},0) = 0$,

$$\mathbf{f}(\mathbf{r},t) = \int_0^t N_1(\mathbf{r},\tau)\mathbf{E}(\mathbf{r},\tau)d\tau. \quad (A.9)$$

From Equations A.2, A.3, and A.9,

$$J_1(\mathbf{r},t) = \varepsilon_0 \int_0^t \omega_{p1}^2(\mathbf{r},\tau)E(\mathbf{r},\tau)d\tau, \tag{A.10}$$

where

$$\omega_{p1}^2(\mathbf{r},t) = \frac{q^2 N_1(\mathbf{r},t)}{m\varepsilon_0}. \tag{A.11}$$

The current density $J_0(\mathbf{r},t)$ due to the steady density $N_0(\mathbf{r},t)$ is easily computed since the initial velocity at $t = 0$ of all the electrons is the same and is given by $v_0(\mathbf{r},0)$

$$\begin{aligned}
J_0(\mathbf{r},t) &= -qN_0(\mathbf{r})v_0(\mathbf{r},t) \\
&= \frac{q^2 N_0(\mathbf{r})}{m} \int_0^t E(\mathbf{r},\tau)d\tau - qN_0(\mathbf{r})v_0(\mathbf{r},0) \\
&= \varepsilon_0 \omega_{p0}^2(\mathbf{r}) \int_0^t E(\mathbf{r},\tau)d\tau + J_0(\mathbf{r},0),
\end{aligned} \tag{A.12}$$

where

$$J_0(\mathbf{r},0) = -qN_0(\mathbf{r})v_0(\mathbf{r},0). \tag{A.13}$$

By adding Equations A.12 and A.10, the total current $J(\mathbf{r},t)$ is obtained:

$$J(\mathbf{r},t) = \varepsilon_0 \int_0^t \omega_p^2(\mathbf{r},\tau)E(\mathbf{r},\tau)d\tau + J(\mathbf{r},0), \tag{A.14}$$

where

$$\omega_p^2(\mathbf{r},t) = \omega_{p0}^2(\mathbf{r}) + \omega_{p1}^2(\mathbf{r},t), \tag{A.15}$$

$$J(\mathbf{r},0) = J_0(\mathbf{r},0). \tag{A.16}$$

Appendix B
Damping Rates of Waves in a Switched Magnetoplasma Medium: Longitudinal Propagation*

Dikshitulu K. Kalluri, University of Massachusetts, and Venkata R. Goteti, Tuskegee University, AL

Abstract — The interaction of a circularly polarized electromagnetic wave with a switched-on magnetoplasma medium is considered. A static magnetic field in the direction of propagation is assumed to be present, resulting in longitudinal propagation. The incident wave splits into three waves whose frequencies are different from that of the incident wave. It is shown that these waves ultimately damp out if the plasma is even slightly lossy. The damping of the waves is interpreted in terms of their attenuation with distance and decay with time as they propagate in the lossy plasma. The attenuation length and decay-time constants of the waves are obtained and their dependence on the incident wave frequency and the gyrofrequency is examined. Optimum parameters for an experiment to detect these waves are suggested.

B.1 INTRODUCTION

The study of the interaction between electromagnetic waves and plasmas is of considerable interest.[1-3] The real-life plasma may be bounded, having time-varying and space-varying (inhomogeneous) parameters. The general problem when all the features are simultaneously present has no analytic solution. At best, one can obtain a numerical solution. In some problems it is possible to neglect one or more of the features and concentrate on the dominant parameter.

Shock waves and controlled-fusion containment experiments are typical examples of time-varying plasmas.[1] The effect of a rapid rise of the electron density to a peak value can be modeled as a sudden creation of a plasma medium; i.e., the plasma medium is suddenly switched on at $t = 0$, creating a temporal discontinuity in the properties of the medium.

* © 1990 IEEE. Reprinted with permission from Kalluri, D. K. and Goteti, V. R., *IEEE Transactions on Plasma Science*, 18, 797–801, 1990.

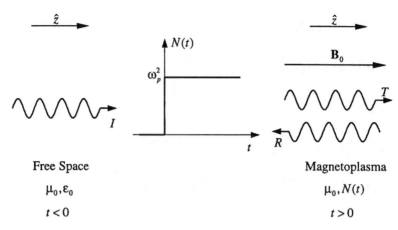

FIGURE B.1   Geometry of the problem. Magnetoplasma medium is switched on at time $t = 0$.

Kalluri[4] has discussed the reflection of a traveling wave when an isotropic plasma of plasma frequency ω_p is switched on only over the $z > 0$ half-space. He has shown that the reflected field in the free space comprises two components, A and B. The A component is due to reflection at the spatial discontinuity at $z = 0$. Its frequency is the same as that of the incident wave. The temporal discontinuity gives rise to waves, called the B waves, and the one propagating in the negative z-direction undergoes partial transmission into free space. This is designated as the B component. Its frequency is different from that of the incident wave. Further, it is shown that the B component will be damped out even if the plasma is only slightly lossy. The B component is shown[3] to carry a maximum 3.7% of the power in the incident wave when the incident wave frequency ω is 0.35 ω_p. For this value of the incident wave frequency, the damping time constant is shown[4] to be 224.5/ω_p s when the collision frequency v is 1% of ω_p. The B component is thus weak and is damped out rapidly.

Kalluri[5] has studied the effect on a traveling wave of switching on an unbounded magnetoplasma medium. The direction of propagation of the incident (circularly polarized) wave is taken to be the same as that of the static magnetic field (longitudinal propagation[6]). He has shown that the original wave splits into three B waves. One of these waves is strongly influenced by the strength of the static magnetic field and carries a significant portion of the incident power for a value of $\omega = 0.67\omega_p$ and the gyrofrequency $\omega_b = 2\omega_p$. If this wave lasts long enough, there is a good chance of detecting its presence, which is motivation to investigate the damping rates of these waves. The results given in Refernce 5 and the results obtained in this appendix are used in suggesting a suitable experiment to detect these waves.

B.2   FREQUENCIES AND VELOCITIES OF THE NEW WAVES

Figure 1 shows the geometry of the problem. Initially, for time $t < 0$, a circularly polarized plane wave is assumed to be propagating in free space in the positive z-direction. The incident wave is designated by

$$\bar{e}(z,t) = (\hat{x} \mp j\hat{y})E_0 \exp\left(j(\omega_0 t - k_0 z)\right) \quad \text{and} \tag{B.1}$$

$$\bar{h}(z,t) = (\pm j\hat{x} + \hat{y})H_0 \exp\left(j(\omega_0 t - k_0 z)\right). \tag{B.2}$$

In Equations B.1 and B.2, the upper sign corresponds to right circular polarization, designated as R incidence, and the lower to left circular polarization designated as L incidence. This convention will be followed throughout the rest of this appendix. Also,

$$H_0 = \frac{E_0}{\eta_0}, \quad \eta_0 = \sqrt{\frac{\mu_0}{\varepsilon_0}}, \quad k_0 = \omega_0\sqrt{\mu_0\varepsilon_0}, \tag{B.3}$$

where μ_0, ε_0 are, respectively, the free-space permeability and permittivity.

At time $t = 0$, the entire space is converted to a magnetoplasma medium. The plasma is assumed to consist of a constant number density N_0, together with a collision frequency ν, and a static magnetic field of strength B_0 directed along the positive z-direction. These values correspond to a plasma frequency ω_p (= $\{N_0 q^2/m\varepsilon_0\}^{1/2}$) and a gyrofrequency ω_b (=qB_0/m). The analysis is based on the conservation of the wave number of the plasma waves produced due to the temporal discontinuity.[1-4] The plasma fields can be described by

$$\bar{e}(z,t) = (\hat{x} \mp j\hat{y})e(t)\exp(-jk_0 z), \quad t > 0, \tag{B.4a}$$

$$\bar{h}(z,t) = (\pm j\hat{x} + \hat{y})h(t)\exp(-jk_0 z), \quad t > 0, \tag{B.4b}$$

$$\bar{v}(z,t) = (\hat{x} \mp j\hat{y})v(t)\exp(-jk_0 z), \quad t > 0. \tag{B.4c}$$

In Equation B.4c, v refers to the velocity field. The plasma fields under this condition satisfy the following equations:

$$\frac{d}{dt}(\eta_0 h) = j\omega_0 e, \quad t > 0 \tag{B.5a}$$

$$\frac{d}{dt}(e) = j\omega_0(\eta_0 h) + \frac{N_0 q}{\varepsilon_0}v, \quad t > 0 \tag{B.5b}$$

$$\frac{d}{dt}(v) = \frac{q}{m}e \pm j\omega_0 v - \nu v, \quad t > 0 \tag{B.5c}$$

In the absence of collisions ($\nu = 0$), the real frequencies of the newly created waves can be obtained by replacing (d/dt) by $j\omega$ in Equation B.5. This will yield the cubic given in Equation B.6:

$$D(\omega) = \omega^3 \mp \omega_b \omega^2 - \left(\omega_0^2 + \omega_p^2\right)\omega \pm \omega_0^2 \omega_b = 0. \tag{B.6}$$

Positive values of the frequencies in Equation B.6 correspond to the frequencies of the transmitted waves and negative values to the frequencies of the reflected waves. A double subscript notation is used wherever necessary to indicate the nature of circular polarization (first subscript) of the incident wave and to indicate the wave being considered (second subscript). The three waves were studied,[5] and it was found that

1. For R incidence, one of the three waves propagates in the negative z-direction; this wave has a frequency ω_{R2}.
2. For L incidence, two of the waves propagate in the negative z-direction; these waves have frequencies ω_{L2} and ω_{L3}.

The group velocity u_{gn} of each of the newly created waves is obtained as

$$\frac{u_{gn}}{c} = \left[\frac{d\omega_n}{d\omega_0}\right] = \frac{2\omega_0\left(\omega_n \mp \omega_b\right)}{3\omega_n^2 \mp 2\omega_n\omega_b - \omega_0^2 - \omega_p^2}, \quad n = 1, 2, 3. \tag{B.7}$$

The velocity of energy transport u_{en} is defined as the ratio of the time-averaged Poynting vector to the total energy density. The total energy density comprises the energy stored in the electric field, the energy stored in the magnetic field, and the kinetic energy of the moving electrons. The energy velocity represents the velocity at which energy is carried by an electromagnetic wave. The energy velocities of the three waves under consideration are given by

$$\frac{u_{en}}{c} = \frac{2\omega_0\omega_n\left(\omega_n \mp \omega_b\right)^2}{\left[\left(\omega_n \mp \omega_b\right)^2\left(\omega_n^2 + \omega_0^2\right) + \omega_n^2\omega_p^2\right]}, \quad n = 1, 2, 3. \tag{B.8}$$

Equation B.8, when simplified with the help of Equation B.6, reduces to Equation B.7. Thus, it can be concluded that the group velocity of the waves adequately characterizes the velocity of energy transfer by the three waves.

From these equations and the $\omega - \beta$ diagrams discussed by Kalluri,[5] the following conclusions are drawn with regard to the dependence of the energy velocity with the gyrofrequency:

1. The energy velocity decreases for the wave having frequency ω_{R1} or ω_{L2}.
2. The energy velocity increases for the wave having frequency ω_{R2} or ω_{L1}.
3. The energy velocity increases for the wave having frequency ω_{R3} or ω_{L3}.

These conclusions are useful in explaining the variation of the damping constants of the waves with the gyrofrequency and the incident wave frequency (Figures B.2 and B.3).

B.3 DAMPING RATES OF THE NEW WAVES

In the ideal case of a lossless plasma, the B waves continue to propagate at all times. They ultimately damp out if the magnetoplasma is even slightly lossy. The possibility of detection of the B waves before they damp out in a lossy magnetoplasma is investigated in this section. To this end, the decay time constant and the attenuation length of the B waves will be obtained. The plasma is considered to be slightly lossy ($v \ll \omega_p$) so that approximation methods can be used for obtaining these parameters.

When collisions are included, each of the three frequencies will be associated with a damping term, thus forming a complex frequency. Therefore, in order to obtain the complex frequencies, it is necessary to replace (d/dt) in Equation B.5 by the complex frequency variable s. The characteristic equation $D(s)$ is obtained from Equation B.5:

$$D(s) = s^3 + s^2\left(v \mp j\omega_b\right) + s\left(\omega_0 + \omega_p^2\right) + \omega_0^2\left(v \mp j\omega_b\right) = 0. \qquad (B.9)$$

The imaginary parts of the roots of $D(s)$ give the real frequencies while their real parts indicate the nature of their exponential decay with time. Taylor's series expansion of s about $v = 0$ gives the relation:

$$s(v) = [s]_{v=0} + v\left[\frac{ds}{dv}\right]_{v=0}. \qquad (B.10)$$

The first term on the right-hand side of Equation B.10 gives the frequency (Equation B.6), and the second term gives the associated damping factor. The three complex roots of Equation B.9 are therefore given by

$$s_n = \sigma_n + j\omega_n, \qquad n = 1, 2, 3, \qquad (B.11a)$$

$$\sigma_n = v\left[\frac{ds}{dv}\right]_{v=0, s=j\omega_n}, \qquad n = 1, 2, 3. \qquad (B.11b)$$

The decay time constants of the three waves are

$$t_{pn} = -\frac{1}{\sigma_n} = \frac{\left(3\omega_n^2 \mp 2\omega_n\omega_b - \omega_0^2 - \omega_p^2\right)}{v\left(\omega_n^2 - \omega_0^2\right)}, \qquad n = 1, 2, 3. \qquad (B.12a)$$

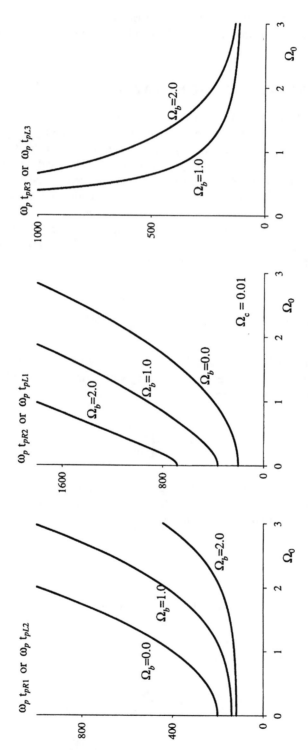

FIGURE B.2 Decay of the waves with time at a given point in space.

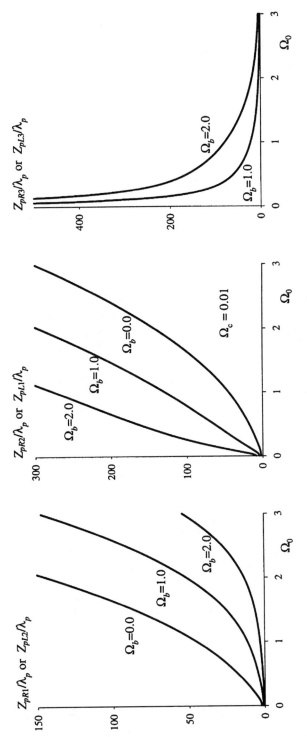

FIGURE B.3   Decay of the waves in space at a given time.

Substitution for ω_b from Equation B.6 gives

$$t_{pn} = \frac{1}{v}\left[1 + \frac{\omega_p^2}{\left(\omega_n^2 - \omega_0^2\right)} + \frac{2\omega_p^2\omega_0^2}{\left(\omega_n^2 - \omega_0^2\right)^2}\right], \quad n = 1, 2, 3. \tag{B.12b}$$

The decay time constants for R incidence and L incidence are related:

$$t_{pR1} = t_{pL2}, \tag{B.13a}$$

$$t_{pR2} = t_{pL1}, \tag{B.13b}$$

$$t_{pR3} = t_{pL3}. \tag{B.13c}$$

The time constant associated with each frequency can be interpreted as the time in which the amplitude of the wave reduces to 36.8% of the initial value.

It is possible to interpret this decay in the time domain as an equivalent attenuation with distance in the space domain. The concept of energy velocity is used in this context. The products of the time constants given in Equation B.13 with the corresponding energy velocities given in Equation B.8 give a measure of distance over which the waves propagate before their amplitudes become 36.8% of their initial values.[4] These are termed the attenuation lengths denoted by Z'_{pn}, and can be written specifically for the three waves:

$$Z'_{pR1} = Z'_{pL2} = t_{pR1}u_{eR1}, \tag{B.14a}$$

$$Z'_{pR2} = Z'_{pL1} = t_{pR2}u_{eR2}, \tag{B.14b}$$

$$Z'_{pR3} = Z'_{pL3} = t_{pR3}u_{eR3}. \tag{B.14c}$$

An alternative way of obtaining the expression for the attenuation length is through the expression for the complex propagation constant $\gamma_n{}^1$ given by Equation B.15:

$$\gamma_n = \alpha_n + j\beta_n = j\mu_n\frac{\omega_n}{c}, \quad n = 1, 2, 3, \tag{B.15a}$$

$$\mu_n^2 = 1 - \frac{\omega_p^2}{\omega_n\left(\omega_n \mp \omega_b - jv\right)}, \quad n = 1, 2, 3. \tag{B.15b}$$

The space attenuation of the plasma waves can be interpreted in terms of the attenuation length constant Z_p given by

$$Z_{pn} = \frac{1}{\alpha_n}, \quad n = 1, 2, 3. \tag{B.16}$$

For the three plasma waves under consideration, the attenuation lengths are given by

$$Z_{pR1} = Z_{pL2} = \frac{1}{\alpha_{R1}}, \tag{B.17a}$$

$$Z_{pR2} = Z_{pL1} = \frac{1}{\alpha_{R2}}, \tag{B.17b}$$

$$Z_{pR3} = Z_{pL3} = \frac{1}{\alpha_{R3}}. \tag{B.17c}$$

The attenuation lengths calculated from Equation B.14 are found to be very close to those obtained from Equation B.17 when the collision frequency is small. The decay time constants and attenuation lengths are computed for various values of the incident wave frequency and the gyrofrequency. The results are discussed in the following.

B.4   NUMERICAL RESULTS

In this analysis, the frequencies are normalized with respect to the plasma frequency. Thus,

$$\text{Frequency of the incident wave:} \quad \Omega_0 = \omega_0/\omega_p \tag{B.18a}$$

$$\text{Gyrofrequency:} \quad \Omega_b = \omega_b/\omega_p \tag{B.18b}$$

$$\text{Collision frequency:} \quad \Omega_c = \nu/\omega_p \tag{B.18c}$$

$$\text{Frequency of the plasma waves:} \quad \Omega_n = \omega_n/\omega_p, \quad n = 1, 2, 3. \tag{B.18d}$$

Figure B.2 shows the variation of the damping time constant of the plasma waves for a collision frequency of $\Omega_c = 0.01$. In these plots the values of $\omega_p t_p$ are plotted as functions of Ω_0. Figure B.2a shows the damping rate of the waves having frequency Ω_{R1}. The same plot holds for the wave having frequency Ω_{L2}. For all values of the incident wave frequency, this wave decays faster as the strength of the static magnetic field is increased. The decay is more pronounced for $\Omega_0 > 1.0$ than for lower frequencies. For lower values of Ω_0, the static magnetic field has little effect (when $\Omega_0 = 0.5$, $\omega_p t_{pR1}$ is 150 for $\Omega_b = 1.0$, while it is 120 for $\Omega_b = 2.0$). Also, for a given Ω_b, high values of Ω_0 result in longer persistence of this wave in the plasma.

Figure 2b shows the decay rates of the wave having frequency Ω_{R2}. The behavior of this wave with Ω_b is different from that of the first wave. As Ω_b is reduced, this wave decays faster. For a specific Ω_b, as Ω_0 is increased, this wave persists longer in the plasma. Figure B.2c shows the time domain decay of the third wave. For a given value of the incident wave frequency, an increase in the value of Ω_b results in a longer time of persistence of this wave. In the region where $\Omega_0 < 1.0$, the rate of decay of this wave with Ω_0 is much larger than its rate of decay for higher values of Ω_0. Figure B.3 shows the normalized attenuation lengths of the plasma waves in the space domain for a normalized collision frequency of $\Omega_c = 0.01$. Here, the attenuation length Z_p is normalized with respect to $\lambda_p (= 2\pi c/\omega_p)$, which is the wavelength in free space corresponding to plasma frequency. In Figure B.3a is shown the normalized attenuation length of the wave having frequency Ω_{R1} (or Ω_{L2}). Z_{pR1}/λ_p starts at a small value and increases to a large value with Ω_0 for any specific Ω_b.

The stronger the imposed magnetic field, the shorter the distance of propagation of this wave. Figure B.3b shows the space behavior of the wave having frequency Ω_{R2} (or Ω_{L1}). The stronger the imposed magnetic field, the longer the distance of propagation of this wave in the plasma. Figure B.3c shows the attenuation of the third wave having frequency Ω_{R3} (or Ω_{L3}). This wave is not present in the isotropic case and is the distinguishing feature of longitudinal propagation in a switched-on plasma medium. A small increase in Ω_0 brings about an appreciable reduction in the attenuation length of this wave when ω_0 is less than the plasma frequency ($\Omega_0 < 1.0$). When Ω_0 tends to infinity, the attenuation length of this wave becomes zero. When the collision frequency v is small, the attenuation length of the waves can be shown to be approximately given by

$$\frac{Z_{pn}}{\lambda_p} \cong \frac{1}{\pi\Omega_c} \frac{\left(\Omega_n \mp \Omega_b\right)^2 \Omega_0}{\Omega_n}, \quad n = 1, 2, 3. \tag{B.19}$$

The damping constants, t_{pn} and Z_{pn}, increase with the energy velocity (group velocity). This is the physical explanation for the variation of the damping constants with Ω_b given in Figures B.2 and B.3.

B.5   SUGGESTED EXPERIMENT

The characteristic wave, the third wave in the case of R incidence, carries considerable power[5] and lasts long (Figure B.3c) for a value of $\Omega_0 = 0.67$ and $\Omega_b = 2.0$. Based on the assumption that the plasma can be considered as suddenly created if the time period of the incident wave is at least 100 times the rise time t_0 of the plasma electron density, the incident wave frequency is given by $\omega_0 = (2\pi/100t_0)$. The other frequency parameters are, therefore, $\omega_p = (3\pi/100t_0)$, $\omega_b = (6\pi/100t_0)$, and $v = 3\pi/10,000t_0)$ for $\Omega_c = 0.01$. The decay time constant of the wave is found to be $t_p = 10,080t_0$. Table B.1 shows the required values of the parameters for two values of t_0.

TABLE B.1
Optimum Parameters for a Suitable Experiment

Rise Time, t_0	Incident Frequency, $\omega_0/2\pi$	Plasma Frequency, $\omega_p/2\pi$	Gyrofrequency, $\omega_b/2\pi$	Collision Frequency, $\nu/2\pi$	Wave to Be Observed (Third Wave)	
					Frequency $\omega_3/2\pi$	Time Decay Constant, t_p
1 μs	10 kHz	15 kHz	30 kHz	150 s^{-1}	6.37 kHz	10 ms
1 ns	10 MHz	15 kHz	30 MHz	150×10^3 s^{-1}	6.37 MHz	10 μs

B.6   CONCLUSIONS

The sudden imposition of a magnetoplasma medium on a traveling wave causes the creation of three new waves having frequencies different from that of the incident wave. Two of these waves are transmitted, and one is a reflected wave when the incident wave has right-hand circular polarization. The situation reverses when the incident wave has left-hand circular polarization. The energy velocity of each wave is the same as its group velocity ($\nu = 0$). The creation of the third wave is the distinguishing feature of the longitudinal propagation.

The attenuation of the B waves in the case of the imposition of a lossy magnetoplasma medium can be estimated through the computation of their decay time constants. The energy velocity can be used to relate the decay time constant to the attenuation length. An alternative means of estimating the attenuation length is through consideration of the complex propagation constant computed at the new frequencies.

REFERENCES

1. Heald, M. A. and Wharton, C. B., *Plasma Diagnostics with Microwaves*, Wiley, New York, 1965.
2. Auld, B. A., Collins, J. H., and Zapp, H. R., Signal processing in a nonperiodically time-varying magnetoelastic medium, *Proc. IEEE*, 56, 258, 1968.
3. Jiang, C. L., Wave propagation and dipole radiation in a suddenly created plasma, *IEEE Trans. Antennas Propag.*, AP-23, 83, 1975.
4. Kalluri, D. K., On reflection from a suddenly created plasma half-space: transient solution, *IEEE Trans. Plasma Sci.*, 16, 11, 1988.
5. Kalluri, D. K., Effect of switching a magnetoplasma medium on a traveling wave: longitudinal propagation, *IEEE Trans. Antennas Propag.*, AP-37, 1638, 1989.
6. Booker, H. G., *Cold Plasma Waves*, Kluwer, Hingham, MA, 1984.

Appendix C
Wave Propagation in a Switched Magnetoplasma Medium: Transverse Propagation*

Venkata R. Goteti and Dikshitulu K. Kalluri

Abstract — The interaction of electromagnetic radiation with a time-switched magnetoplasma is considered. It is shown that for the case of the incident electric field normal to the static magnetic field, the original wave splits into four new extraordinary waves whose frequencies are different from the incident wave frequency. Two of these new waves are transmitted waves and two are reflected waves. The characteristics of these waves (frequencies, power carried by them, and their damping rates) are discussed.

C.1 INTRODUCTION

The study of interaction between electromagnetic radiation and plasmas having time-varying parameters is of considerable interest.[1-6] The interaction is characterized by the creation of new waves of frequencies different from that of the incident wave. The nature of these new waves (amplitudes, frequencies, power carried) can be studied through a theoretical model of a suddenly created plasma. In this model it is assumed that a plane wave of frequency ω_0 is traveling in free space. At time $t = 0$, the free-electron density in the medium increases suddenly from zero to some constant value N_0, i.e., a plasma of frequency ω_p, where

$$\omega_p = \left(N_0 q^2 / m\varepsilon_0\right)^{1/2} \tag{C.1}$$

is switched on at $t = 0$. In Equation C.1, q is the numerical value of the charge of an electron, m is the mass of an electron, and ε_0 is the permittivity of free space.

In the absence of a static magnetic field, the created plasma behaves like an isotropic medium. It is known[3] that in the isotropic case two new waves are created of frequencies

$$\omega_{1,2} = \pm\left(\omega_0^2 + \omega_p^2\right)^{1/2}. \tag{C.2}$$

* © 1990 American Geophysical Union. Reprinted with permission from Goteti, V. R. and Kalluri, D. K., *Radio Science*, 25, 61–72, 1990.

In Equation C.2 the negative value for the frequency indicates a reflected wave.

In the presence of the static magnetic field, the switched medium is anisotropic. The case of longitudinal propagation (the static magnetic field in the direction of wave propagation and the incident wave circularly polarized) was investigated by Kalluri[6] in 1989.

This appendix investigates the effect of switching on a magnetoplasma medium for the case of transverse propagation. The incident wave (the wave existing prior to switching the medium) is assumed to be linearly polarized and propagating in the positive z-direction. The static magnetic field is assumed to be along the positive y-direction (transverse propagation).[7] The case of electric field along the y-direction results in ordinary wave propagation. It is the same as the isotropic case and has already been discussed in detail by one of the authors.[5] The case of the incident electric field along the x-axis results in the extraordinary mode[7] of propagation and is the subject of this appendix.

In Section C.2 the basic nature of the interaction is studied in terms of the frequencies, amplitudes of the fields, and the power carried by the new waves. For this purpose the switched medium is assumed to occupy the entire space for $t > 0$. The fields of these waves will be attenuated if the plasma is lossy. The damping rates in time and space of these waves are examined in Section C.3. A more realistic situation of switching on a finite extent magnetoplasma is modeled in Section C.4 by considering the switched medium as a magnetoplasma half-space. The steady-state solution to this problem is developed along the lines of Kalluri,[5] who considered the isotropic case. The transient solution to the switched lossy magnetoplasma half-space problem is developed in Section C.5 on the lines of Kalluri.[5] Numerical Laplace inversion of the solution confirms the damping rates of Section C.3 and the steady-state solution of Section C.4. In Section C.6 the numerical results are discussed. In the following analysis, ion motion is neglected (radio approximation).

C.2   FREQUENCIES AND POWER CONTENT OF THE NEW WAVES

The geometry of the problem is shown in Figure C.1a. Initially, a plane wave is assumed to be propagating in free space in the positive z-direction. Thus, for $t < 0$ the fields are described by

$$\mathbf{e}(z,t) = \hat{x} E_0 \exp\left[j\left(\omega_0 t - k_0 z\right)\right], \quad t < 0, \tag{C.3}$$

$$\mathbf{h}(z,t) = \hat{y} H_0 \exp\left[j\left(\omega_0 t - k_0 z\right)\right], \quad t < 0, \tag{C.4}$$

where ω_0 is the incident wave frequency, k_0 is the free-space wave number, μ_0 and ε_0 are the free-space parameters, and

$$H_0 = E_0 / \eta_0, \quad \eta_0 = \sqrt{\mu_0 / \varepsilon_0}. \tag{C.5}$$

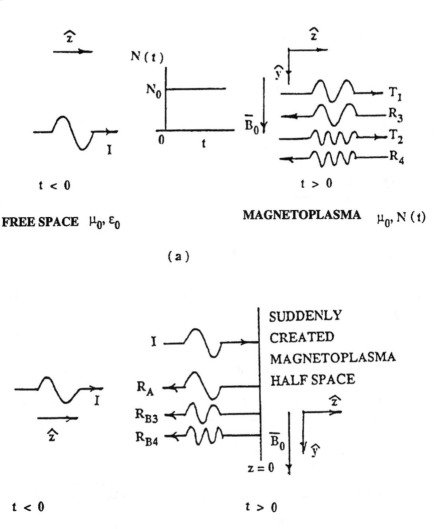

FIGURE C.1 Geometry of the problem: switching of (a) an unbounded magnetoplasma medium and (b) a magnetoplasma half-space. The electric field of the incident wave is normal to the static magnetic field B_0.

At time $t = 0$, the entire space is converted to a magnetoplasma. The plasma is characterized by a constant number density N_0 and an imposed static magnetic field of strength B_0 along the y-direction. The fields satisfy the following equations under this condition (for $t > 0$):

$$\nabla \times \mathbf{e} = -\mu_0 \frac{\partial \mathbf{h}}{\partial t}, \qquad t > 0, \tag{C.6}$$

$$\nabla \times \mathbf{h} = \varepsilon_0 \frac{\partial \mathbf{e}}{\partial t} - N_0 q \mathbf{v}, \qquad t > 0, \tag{C.7}$$

$$\frac{\partial \mathbf{v}}{\partial t} = -\frac{q}{m} \mathbf{e} - \frac{q}{m} \mathbf{v} \times \mathbf{B}_0, \qquad t > 0, \tag{C.8}$$

where **v** denotes the velocity field.

The waves in the plasma will be extraordinary waves.[7] So the plasma fields can be written as

$$\mathbf{e} = \hat{x} e_x + \hat{z} e_z, \tag{C.9a}$$

$$\mathbf{h} = \hat{y} h_y, \tag{C.9b}$$

$$\mathbf{v} = \hat{x} v_x + \hat{z} v_z. \tag{C.9c}$$

It is now possible to express Equations C.6 through C.8 as partial differential equations in z and t variables. With regard to the wave propagation in time-varying media, it is shown in the works of Fante,[2] Jiang,[3] and Kalluri[5] that the wave number is conserved. Therefore, the z variation of the fields can be written in the form

$$f(z,t) = f(t) \exp(-jk_0 z). \tag{C.10}$$

Based on Equation C.10, the field Equations C.6 through C.9 can be written as

$$\frac{d}{dt}(\eta_0 h_0) = j\omega_0 e_x, \tag{C.11a}$$

$$\frac{d}{dt}(e_x) = j\omega_0(\eta_0 h_y) + \frac{N_0 q}{\varepsilon_0} v_x, \tag{C.11b}$$

$$\frac{d}{dt}(e_z) = \frac{N_0 q}{\varepsilon_0} v_z, \tag{C.11c}$$

$$\frac{d}{dt}(v_x) = -\frac{q}{m} e_x + \omega_b v_z, \tag{C.11d}$$

$$\frac{d}{dt}(v_z) = -\frac{q}{m} e_z - \omega_b v_x, \tag{C.11e}$$

where ω_b is the gyrofrequency,

$$\omega_b = \frac{qB_0}{m}. \tag{C.11f}$$

Solution of Equation C.11 with appropriate initial conditions gives the time domain description of the fields for $t = 0$.

It is assumed that the newly created electrons are at stationary $t = 0^3$ and that they are set in motion only for $t > 0$. The free-current density in the medium thus remains at zero value over the time discontinuity. Therefore, the tangential electric and magnetic fields are continuous over $t = 0$. Further, it is assumed that the longitudinal component of the electric field, e_z, is present only for $t > 0$. The initial conditions are thus

$$v_x(z,0) = 0, \tag{C.12a}$$

$$v_z(z,0) = 0, \tag{C.12b}$$

$$e_x(z,0^-) = e_x(z,0^+) = E_0 \exp(-jk_0 z), \tag{C.12c}$$

$$h_y(z,0^-) = h_y(z,0^+) = H_0 \exp(-jk_0 z), \tag{C.12d}$$

$$e_z(z,0) = 0. \tag{C.12e}$$

Equations C.11 are solved through the use of Laplace transforms together with the initial conditions specified in Equations C.12. The Laplace transform of $f(t)$ is defined as

$$F(s) = \int_0^\infty f(t)e^{-st} dt. \tag{C.13}$$

The Laplace transforms of the field quantities are obtained as

$$E_x(z,s) = \frac{N_1(s)}{D(s)} E_0 \exp(-jk_0 z), \tag{C.14a}$$

$$E_z(z,s) = \frac{N_2(s)}{sD(s)} E_0 \exp(-jk_0 z), \tag{C.14b}$$

$$H_y(z,s) = \frac{1}{\eta_0}\left(j\omega_0 \frac{N_1(s)}{sD(s)} + \frac{1}{s}\right) E_0 \exp(-jk_0 z), \tag{C.14c}$$

where

$$N_1(s) = \left(s + j\omega_0\right)\left(s^2 + \omega_b^2 + \omega_p^2\right),$$ (C.15a)

$$N_2(s) = \left(s + j\omega_0\right)\omega_b\omega_p^2,$$ (C.15b)

$$D(s) = \left(s^2 + \omega_0^2\right)\left(s^2 + \omega_b^2 + \omega_p^2\right) + \omega_p^2\left(s^2 + \omega_p^2\right).$$ (C.15c)

Observation of Equations C.14 and C.15 shows that the poles of $E_x(z,s)$, $E_z(z,s)$, and $H_y(z,s)$ are all on the $j\omega$ axis. They are the zeros of $D(s)$. By setting $s = j\omega$ in Equation C.15, the zeros of $D(s)$ can be obtained as the roots of the equation

$$\omega^4 - \omega^2\left(\omega_0^2 + \omega_b^2 + 2\omega_p^2\right) + \left[\omega_p^4 + \omega_0^2\left(\omega_b^2 + \omega_p^2\right)\right] = 0.$$ (C.16)

Equation C.16 is a quadratic in ω^2 and can be readily solved to give four frequencies:

$$\omega_1 = \left(A + \sqrt{A^2 - B}\right)^{1/2},$$ (C.17a)

$$\omega_2 = \left(A - \sqrt{A^2 - B}\right)^{1/2},$$ (C.17b)

$$\omega_3 = -\omega_1,$$ (C.17c)

$$\omega_4 = -\omega_2,$$ (C.17d)

where

$$A = \left(\omega_0^2 + \omega_b^2 + 2\omega_p^2\right)\Big/2,$$ (C.18a)

$$B = \omega_p^4 + \omega_0^2\left(\omega_b^2 + \omega_p^2\right).$$ (C.18b)

The important conclusion that can be drawn from Equations C.17 and C.18 is that ω_1 and ω_2 are real and positive and are the frequencies of the transmitted waves. The frequencies ω_3 and ω_4 are the frequencies of the corresponding reflected waves. Thus, the incident wave of frequency ω_0 splits into four waves because of the imposition of the magnetoplasma. All these four waves will have the same wave number as that of the incident wave. The variation of these frequencies with the incident wave frequency is discussed in Section C.6 (Figure C.2).

By referring again to Equations C.14b and C.14c, it is evident that in addition to these four waves there is a purely space-varying component present in e_z and h_y. This corresponds to the residue of the pole at the origin of $E_z(s)$ and $H_y(s)$. The complete expression for the individual frequency components of $e_x(z,t)$, $e_z(z,t)$, and $h_y(z,t)$ can be obtained as

$$\frac{e_{xn}(z,t)}{E_0} = \frac{(\omega_n + \omega_0)(\omega_n^2 - \omega_p^2 - \omega_b^2)\exp\left[j(\omega_n t - k_0 z)\right]}{\displaystyle\prod_{m=1,m\neq n}^{4}(\omega_n - \omega_m)}, \quad n=1,2,3,4, \quad (\text{C.19})$$

$$\frac{e_{zn}(z,t)}{E_0} = j\frac{(\omega_n + \omega_0)\omega_b\omega_p^2\exp\left[j(\omega_n t - k_0 z)\right]}{\omega_n\displaystyle\prod_{m=1,m\neq n}^{4}(\omega_n - \omega_m)}, \quad n=1,2,3,4, \quad (\text{C.20a})$$

$$\frac{e_{z0}(z,t)}{E_0} = j\frac{\omega_0\omega_b\omega_p^2}{\omega_1\omega_2\omega_3\omega_4}\exp\left(-jk_0 z\right), \quad (\text{C.20b})$$

$$h_{yn}(z,t) = \frac{1}{\eta_0}\frac{\omega_0}{\omega_n}e_{xn}(z,t), \quad n=1,2,3,4, \quad (\text{C.21a})$$

$$\frac{h_{y0}(z,t)}{H_0} = \frac{\omega_p^4}{\omega_1\omega_2\omega_3\omega_4}\exp\left(-jk_0 z\right). \quad (\text{C.21b})$$

The active power associated with e_x and h_y components is along the direction of propagation. No active power is associated as a result of e_z and h_y component interaction. The relative power content of the four waves expressed as a ratio of the time-averaged Poynting vector for each frequency component to that of the incident wave is given by

$$S_n = \frac{\langle e_{xn}h_{yn}\rangle}{E_0 H_0/2} \quad n=1,2,3,4. \quad (\text{C.22})$$

The relative power content of the waves with frequencies ω_1 and ω_2 is found to be positive, confirming transmission, and that for ω_3 and ω_4 is found to be negative, confirming reflection. The variation of the power content of these waves with the incident wave frequency is discussed below in Section C.6 (Figure 3).

C.3   DAMPING RATES FOR THE NEW WAVES

The effects of switching on of a lossy plasma on wave propagation are examined in this section by introducing a collision frequency v in the plasma field equations. The momentum equation (Equation C.8) gets modified as

$$\frac{d\mathbf{v}}{dt} = -\frac{q}{m}\mathbf{e} - \frac{q}{m}\mathbf{v} \times \mathbf{B}_0 - v\mathbf{v}. \tag{C.23}$$

Analysis can be carried out as in Section C.2, and the expression for the fields can be obtained. These are given by Equations C.14 and C.15, where ω_p is replaced by $C(s)$ and

$$C(s) = \frac{s\omega_p^2\left[s(s+v)+\omega_p^2\right]}{\left[s(s+v)^2+\omega_p^2(s+v)+\omega_b^2 s\right]}. \tag{C.24}$$

After simplifying, one obtains

$$E_x(z,s) = E_x(s)E_0 \exp(-jk_0 z), \tag{C.25a}$$

$$E_x(s) = \frac{NR(s)}{DR(s)}, \tag{C.25b}$$

where

$$NR(s) = (s+j\omega_0)\left[s(s+v)^2 + \omega_p^2(s+v) + \omega_b^2 s\right], \tag{C.26a}$$

$$DR(s) = s^5 + a_4 s^4 + a_3 s^3 + a_2 s^2 + a_1 s + a_0, \tag{C.26b}$$

$$a_4 = 2v, \tag{C.26c}$$

$$a_3 = \left(\omega_0^2 + v^2 + \omega_b^2 + 2\omega_p^2\right), \tag{C.26d}$$

$$a_2 = 2v\left(\omega_0^2 + \omega_p^2\right), \tag{C.26e}$$

$$a_1 = \omega_0^2\left(v^2 + \omega_b^2 + \omega_p^2\right) + \omega_p^4, \tag{C.26f}$$

$$a_0 = v\omega_0^2\omega_p^2. \tag{C.26g}$$

Further, it is possible to write Equation C.26b as

$$DR(s) = (s + s_0)\left[(s + s_1)^2 + \omega_{11}^2\right]\left[(s + s_2)^2 + \omega_{21}^2\right]. \tag{C.27}$$

The basis for the assumption of a solution of this form is the fact that, because of collisions, the fields produced in the plasma as a result of a temporal discontinuity die out ultimately.[3,5] Therefore, the introduction of a collision frequency term contributes to the following changes in the nature of the plasma waves. The first is a modification of the frequencies from ω_1 to ω_{11} and from ω_2 to ω_{21}. When ν is small, ω_{11} and ω_{21} will be close to ω_1 and ω_2. The second is the decay of these waves. Waves having frequency ω_{11} decay with a time constant of $1/s_1$, and those with ω_{21} frequency have a decay time constant of $1/s_2$. The third is an addition of a pure exponential decay term with a time constant of $1/s_0$. In the absence of collisions this term is not present. Thus, the x component of the electric field can be written as

$$e_x(z,t) = \sum_{n=0}^{4} e_{xn}(t)\exp(-jk_0 z), \tag{C.28a}$$

$$e_{x0}(t) = E_{p0}\exp(-t/t_{p0}), \tag{C.28b}$$

$$e_{xn}(t) = E_{pn}\exp(-t/t_{pn}\exp)(-j\omega_{n1}t), \quad n = 1, 2, 3, 4, \tag{C.28c}$$

where

$$t_{p0} = \frac{1}{s_0}, \tag{C.29a}$$

$$t_{p1} = t_{p3} = \frac{1}{s_1}, \tag{C.29b}$$

$$t_{p2} = t_{p4} = \frac{1}{s_2}, \tag{C.29c}$$

$$\omega_{31} = -\omega_{11}, \tag{C.29d}$$

$$\omega_{41} = -\omega_{21}. \tag{C.29e}$$

In Equation C.28, E_{p0}, E_{p1}, E_{p2}, E_{p3}, and E_{p4} are the residues of $E_x(s)$ given in Equation C.25 at the five poles of $E_x(s)$. For specified values of ω_0, ω_p, ω_b, and ν, the plasma electric field is determined, and the amplitude and decay of each of the five component terms are examined.

While the foregoing analysis is concerned with the time domain decay of the plasma waves, a second way of interpreting the decay is possible.[5] This concerns attenuation of the waves in the space domain as they propagate in the plasma. The complex propagation constant of the plasma waves is given by[8]

$$\gamma_n = \alpha_n + j\beta_n = j\mu_n \frac{\omega_n}{c},$$ (C.30a)

where

$$\mu_n^2 = 1 - \frac{\left[\omega_p^2/\omega_n^2\right]}{1-\left(jv/\omega_n\right)-\left(\omega_b^2/\omega_n^2\right)\left[1-\left(jv/\omega_n\right)-\left(\omega_p^2/\omega_n^2\right)\right]^{-1}}.$$ (C.30b)

In this analysis the wave propagation is characterized by the term $\exp(-\gamma_n z)$, indicating thereby that as the nth wave propagates in the plasma, it will be attenuated in space. The space attenuation constant may be defined as

$$Z_{pn} = \frac{1}{\alpha_n}.$$ (C.31)

Thus, it is possible to explain the decay of plasma waves in two ways: in terms of a decay with time by the decay time constants t_{p1}, t_{p2}, t_{p3}, and t_{p4} and in terms of an attenuation with distance by the space attenuation constants Z_{p1}, Z_{p2}, Z_{p3}, and Z_{p4}. The variation of the damping rates with the incident wave frequency is discussed in Section C.6 (Figures C.6 and C.7).

C.4   MAGNETOPLASMA HALF-SPACE: STEADY-STATE SOLUTION

The switched-on magnetoplasma is considered to be confined to the $z > 0$ half-space in this analysis (Figure C.lb). This problem presents many interesting features. It presents a sharp discontinuity in space in addition to a sharp discontinuity in time. Therefore, in addition to an initial value problem in the time domain, it poses a boundary value problem in the space domain. The distinguishing feature of propagation in a time discontinuity and in a space discontinuity lies in the fact that time discontinuity results in the conservation of the wave number whereas space discontinuity results in the conservation of the frequency. Thus, the plasma waves produced as a result of time discontinuity will consist of a set of four waves having a wave number the same as that of the incident wave. Two of these waves will propagate in the positive z-direction, whereas the remaining two propagate in the negative z-direction. Added to these waves will be one transmitted wave and one reflected wave produced as a result of the space discontinuity. These waves will have the same frequency as the incident wave frequency but with a different wave number. In free space ($z < 0$), there will be three reflected waves. One will be a wave having relative amplitude R_A and frequency ω_0 produced by reflection from the $z = 0$ space boundary. The other two waves are due to those produced in the plasma by the time discontinuity and traveling in the negative z-direction. These waves will be partially transmitted into free space when they encounter the space boundary at $z = 0$. If the plasma is lossless, the two waves continue to propagate in steady state.[5] Their frequencies are already designated as ω_3 and ω_4, and the relative amplitudes R_{B3} and R_{B4}, respectively. Therefore, the total free-space reflected field in steady state is given by a superposition of the three waves. In other words, when the plasma is lossless and the incident wave is $E_0 \cos(\omega_0 t - k_0 z)$, the total reflected electric field in steady state at the interface, $z = 0$, can be written as

$$A_{1Rss}(t) = R \cos \omega_0 t - X \sin \omega_0 t + R_{B3} \cos \omega_3 t + R_{B4} \cos \omega_4 t, \qquad \text{(C.32a)}$$

where

$$R_A = R + jX = \frac{\left(\eta_{p0} - \eta_0\right)}{\left(\eta_{p0} + \eta_0\right)}, \qquad \text{(C.32b)}$$

$$\eta_{p0} = \eta_0 \left(\varepsilon_{r0}\right)^{-1/2}, \qquad \text{(C.32c)}$$

$$\varepsilon_{r0} = 1 - \frac{\omega_p^2/\omega_0^2}{1 - \left(\omega_b^2/\omega_0^2\right)\left[1 - \left(\omega_p^2/\omega_0^2\right)\right]^{-1}}, \qquad \text{(C.32d)}$$

$$R_{B3} = R_{T3} T_{S3}, \qquad \text{(C.32e)}$$

$$R_{B4} = R_{T4} T_{S4}, \qquad \text{(C.32f)}$$

$$R_{T3} = \frac{\left(\omega_3 + \omega_0\right)\left(\omega_3^2 - \omega_p^2 - \omega_b^2\right)}{\left(\omega_3 - \omega_1\right)\left(\omega_3 - \omega_2\right)\left(\omega_3 - \omega_4\right)}, \qquad \text{(C.32g)}$$

$$R_{T4} = \frac{\left(\omega_4 + \omega_0\right)\left(\omega_4^2 - \omega_p^2 - \omega_b^2\right)}{\left(\omega_4 - \omega_1\right)\left(\omega_4 - \omega_2\right)\left(\omega_4 - \omega_3\right)}, \qquad \text{(C.32h)}$$

$$T_{Sn} = 2\eta_0 \left(\eta_{pn} + \eta_0\right)^{-1} \qquad n = 3, 4, \qquad \text{(C.32i)}$$

$$\eta_{pn} = \eta_0 \frac{|\omega_n|}{\omega_0} \qquad n = 3, 4. \qquad \text{(C.32j)}$$

In Equation C.32, R_A is the reflection coefficient from free space to plasma; η_{p0} is the plasma intrinsic impedance at frequency ω_0; R_{T3} and R_{T4} are the relative amplitudes of the negatively propagating waves given by Equation C.19 in Section C.2; T_{S3} and T_{S4} are the transmission coefficients from plasma to free space at frequencies ω_3 and ω_4, respectively; and, finally, R_{B3} and R_{B4} are the relative amplitudes of the transmitted parts of R_{T3} and R_{T4} into free space. The frequencies ω_1, ω_2, ω_3, and ω_4 are obtained in Equation C.17. The variation of the various reflection and transmission coefficients (mentioned earlier) with the incident wave frequency is discussed in Section C.6 (Figure C.4). The fields at any distance z from the interface in free space can be obtained by replacing t with $(t+z/c)$ in Equation C.32.

When the plasma is lossy, even slightly, the steady-state reflected field is only that due to R and X in Equation C.32a since R_{B3} and R_{B4} components die out in steady state. To obtain the steady-state fields when the plasma is lossy, it is necessary to include the effect of the collision frequency v in the expression for the dielectric constant, and it is given by[8]

$$\varepsilon_{r0} = 1 - \frac{\omega_p^2/\omega_0^2}{1-\left(jv/\omega_0\right)-\left(\omega_b^2/\omega_0^2\right)\left[1-\left(jv/\omega_0\right)-\left(\omega_p^2/\omega_0^2\right)\right]^{-1}}, \quad v \neq 0. \quad (C.33)$$

C.5 MAGNETOPLASMA HALF-SPACE: TRANSIENT SOLUTION

The transient electric field in the plasma and free space can be obtained by formulating the problem as an initial value problem in the time domain and as a boundary value problem in the space domain. This involves solution of partial differential equations for the fields. The electric field of the incident wave is written as

$$\mathbf{e}(z,t) = \hat{x}\,\mathrm{Re}\left\{E_0 \exp\left[j(\omega_0 t - k_0 z)\right]\right\}, \quad t < 0. \quad (C.34)$$

At time $t = 0$, the entire $z > 0$ half-space is converted to magnetoplasma. The partial differential equations describing the fields can be written as

$$\frac{\partial^2 e_{xf}}{\partial z^2} - \frac{1}{c^2}\frac{\partial^2 e_{xf}}{\partial t^2} = 0, \qquad t > 0, \ z < 0, \qquad (C.35)$$

$$\frac{\partial^2 e_{xp}}{\partial z^2} - \frac{1}{c^2}\frac{\partial^2 e_{xp}}{\partial t^2} + N_0 q\mu_0 \frac{\partial v_{xp}}{\partial t} = 0, \quad z > 0, \ t > 0, \qquad (C.36a)$$

$$\frac{\partial v_{xp}}{\partial t} = -\left(\frac{q}{m}\right)e_{xp} + \omega_b v_{zp}, \qquad t > 0, \ z > 0, \qquad (C.36b)$$

$$\frac{\partial v_{zp}}{\partial t} = -\left(\frac{q}{m}\right)e_{zp} - \omega_b v_{xp}, \qquad t > 0, \quad z > 0, \qquad (C.36c)$$

$$\frac{\partial e_{zp}}{\partial t} = -\left(\frac{N_0 q}{\varepsilon_0}\right)v_{zp}, \qquad t > 0, \quad z > 0. \qquad (C.36d)$$

Equation C.35 describes the free-space electric field e_{xf}, and Equation C.36 describes the fields in the plasma. Taking Laplace transforms of Equations C.35 and C.36 with respect to t and imposing the initial conditions,

$$e_{xf}(z,0) = e_{xp}(z,0) = e(z,0) = E_0 \exp(-jk_0 z), \qquad (C.37a)$$

$$\frac{\partial e_{xf}(z,0)}{\partial t} = \frac{\partial e_{xp}(z,0)}{\partial t} = \frac{\partial e(z,0)}{\partial t} = j\omega_0 E_0 \exp(-jk_0 z), \qquad (C.37b)$$

$$e_{zp}(z,0) = 0, \qquad (C.37c)$$

$$v_{xp}(z,0) = 0, \qquad (C.37d)$$

$$v_{zp}(z,0) = 0, \qquad (C.37e)$$

one obtains

$$\frac{d^2 E_{xf}(z,s)}{dz^2} - q_1^2 E_{xf}(z,s) = -\frac{(s+j\omega_0)}{c^2} E_0 \exp(-jk_0 z), \qquad (C.38)$$

$$\frac{d^2 E_{xf}(z,s)}{dz^2} - q_2^2 E_{xp}(z,s) = -\frac{(s+j\omega_0)}{c^2} E_0 \exp(-jk_0 z), \qquad (C.39)$$

where

$$q_1 = \frac{s}{c}, \qquad (C.40a)$$

$$q_2 = \frac{1}{c}\left(s^2 + \frac{\omega_p^2\left(s^2 + \omega_p^2\right)}{\left(s^2 + \omega_p^2 + \omega_b^2\right)}\right)^{1/2}. \qquad (C.40b)$$

The solution to these ordinary differential equations can be obtained as

$$E_{xf}(z,s) = A_1(s)\exp(q_1 z) + \frac{(s+j\omega_0)}{(s^2+\omega_0^2)} E_0 \exp(-jk_0 z), \qquad (C.41)$$

$$E_{xp}(z,s) = A_2(s)\exp(-q_2 z) + \frac{(s+j\omega_0)}{c^2(q_2^2+k_0^2)} E_0 \exp(-jk_0 z). \qquad (C.42)$$

The second term on the right side of Equation C.41 is the Laplace transform of the incident electric field. The first term should therefore correspond to the reflected field. The undetermined coefficients $A_1(s)$ and $A_2(s)$ can be obtained from the continuity condition of tangential electric and magnetic fields at $z = 0$. Thus,

$$E_{xf}(0,s) = E_{xp}(0,s), \qquad (C.43a)$$

$$\frac{dE_{xf}}{dz}(0,s) = \frac{dE_{xp}}{dz}(0,s). \qquad (C.43b)$$

By using these conditions, $A_1(s)$ can be shown to be given by

$$\frac{A_1(s)}{E_0} = \frac{(q_2 - jk_0)}{(q_1 + q_2)}\left[\frac{(s+j\omega_0)}{c^2(q_2^2+k_0^2)} - \frac{(s+j\omega_0)}{(s^2+\omega_0^2)}\right]. \qquad (C.44)$$

The reflected electric field thus becomes

$$\frac{e_{xR}}{E_0} = \pounds^{-1}\left\{\frac{A_{1R}(s)}{E_0}\exp(q_1 z)\right\}, \qquad (C.45)$$

where A_{1R}, (s). The real part of $A_1(s)$ is given by

$$\frac{A_{1R}(s)}{E_0} = \left[\frac{-\omega_p^2(sq_2 + \omega_0 k_0)(s^2+\omega_p^2)}{(q_1+q_2)(s^2+\omega_0^2)\left[(s^2+\omega_0^2)(s^2+\omega_p^2+\omega_b^2)+\omega_p^2(s^2+\omega_p^2)\right]}\right]. \qquad (C.46)$$

Numerical Laplace inversion of this irrational function is performed, and it is shown that for large t the total reflected field $A_{1Rss}(t)/E_0$ agrees with that obtained from Equation C.32. See Figure C.5 of Section C.6.

When a loss term ν is introduced, it is possible to show that the reflected field $A_{1R}(s)/E_0$ can be obtained by replacing ω_p by ω_{p1} in Equation C.46, where ω_{p1} is given by Equation C.24. Numerical Laplace inversion then gives the reflected transient electric field of the lossy plasma. In steady state this should agree with the expression given by R_A. Numerical computations verified that this is indeed the case.

C.6 NUMERICAL RESULTS AND DISCUSSION

In the following analysis, frequencies are normalized with respect to the plasma frequency. The normalized variables are denoted as

$$\text{Incident wave frequency} \quad \Omega_0 = \omega_0/\omega_p \qquad \text{(C.47a)}$$

$$\text{Gyrofrequency} \quad \Omega_b = \omega_b/\omega_p \qquad \text{(C.47b)}$$

$$\text{Collision frequency} \quad \Omega_c = \nu/\omega_p \qquad \text{(C.47c)}$$

$$\text{Frequency of the new waves} \quad \Omega_n = \omega_n/\omega_p \qquad \text{(C.47d)}$$

In Figure C.2 is shown the variation of the frequency of the plasma waves with the incident wave frequency (Equations C.17 and C.18). The curves for $\Omega_b = 0$ correspond to the isotropic case. The curves for Ω_1 and Ω_2 start at the cutoff frequencies Ω_{C1} and Ω_{C2}, respectively, given by[8]

$$\Omega_{C1,C2} = \sqrt{\left(\frac{\Omega_b}{2}\right)^2 + 1} \pm \frac{\Omega_b}{2}. \qquad \text{(C.48a)}$$

Here, the upper sign refers to Ω_{C1} and the lower sign to Ω_{C2}. For large values of Ω_0, the Ω_1 curves are found to be asymptotic to the 45° line. This indicates that for high frequencies of the incident wave, the plasma will have no effect and the waves propagate with a group velocity equal to the velocity of light in free space. On the other hand, Ω_2 exhibits certain different features. At $\Omega_0 = 1.0$, Ω_2 is 1.0 for all values of Ω_b. For large Ω_0 values, Ω_2 saturates at a value equal to the normalized upper hybrid frequency Ω_{uh}[8] with zero slope. This value is given by

$$\Omega_{uh} = \left(\Omega_b^2 + 1\right)^{1/2}. \qquad \text{(C.48b)}$$

For the isotropic case with $\Omega_b = 0.0$, Ω_1 and Ω_2 follow the relations

$$\Omega_1 = \sqrt{1 + \Omega_0^2}, \qquad \text{(C.48c)}$$

$$\Omega_2 = 1. \qquad \text{(C.48d)}$$

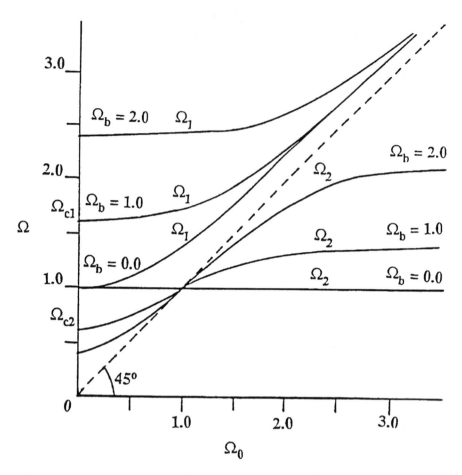

FIGURE C.2 Frequencies of the plasma waves, Ω, vs. incident wave frequency, Ω_0. The results are presented in terms of normalized variables. The horizontal axis is $\Omega_0 = \omega_0/\omega_p$, where ω_0 is the frequency of the incident wave and ω_p is the plasma frequency of the switched medium. The vertical axis is $\Omega = \omega/\omega_p$, where ω is the frequency of the created waves. The branch Ω_n describes the nth created wave. The parameter $\Omega_b = \omega_b/\omega_p$, where ω_b is the gyrofrequency of the switched medium.

Even though Ω_2 appears to be present in the isotropic case, it can be shown that this wave is not excited (see Figures C.3b and d; $S_2 = 0$ and $S_4 = 0$ for $\Omega_b = 0$).

Figure C.3 shows the dependence of the relative power content of the four plasma waves with Ω_0. In Figure C.3a the data are shown for the first transmitted wave, and in Figure C.3c the data are shown for the first reflected wave. Both these waves have the same frequency (Ω_1 or $|\Omega_3|$). The curves in Figures C.3b and d correspond to the second pair of waves with frequency Ω_2 or $|\Omega_4|$. S_1 and S_2 are positive, indicating propagation in the positive z-direction, whereas S_3 and S_4 are negative, indicating propagation in the negative z-direction. When $\Omega_b = 0.0$, only S_1 and S_3 are present. S_1 is observed to increase monotonically from zero to unity for a given Ω_b. For a given Ω_0, an increase in Ω_b results in a reduction in S_1.

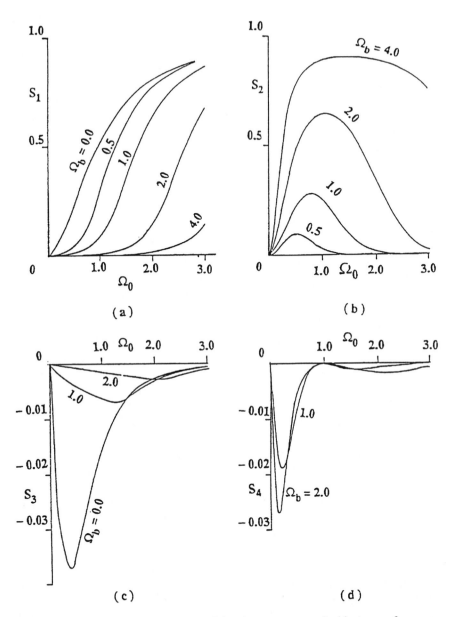

FIGURE C.3 Relative power content of the plasma waves vs. incident wave frequency.

The S_2 curve is observed to start from zero at $\Omega_b = 0$, increase up to a peak value, and finally reduce to zero as Ω_0 increases, for a given Ω_b. An increase in Ω_b increases the value of S_2 for a specific Ω_0. S_2 possesses a peak value of 0.275 for $\Omega_0 = 0.8$ and $\Omega_b = 1.0$, and it is 0.64 at $\Omega_0 = 1.05$ and $\Omega_b = 2.0$.

The relative power content in the first reflected wave is shown by the curve S_3 in Figure C.3c. Its power content is much smaller than that of the first transmitted wave. S_3 has a peak of about 0.036 (about 3.6% only). The stronger the imposed magnetic field, the weaker this wave. The peak power for $\Omega_b = 1.0$ is found to be 0.0068 at $\Omega_0 = 1.2$.

The second reflected wave has a power content S_4 shown in Figure C.3d. S_4 is zero at $\Omega_0 = 1.0$ for all Ω_b. This is because the corresponding electric and magnetic fields are both zero at this frequency. Even though S_4 is small compared with S_2, it is comparable with S_3. S_4 starts from zero at $\Omega_0 = 0.0$, increases to a negative peak value, becomes zero at $\Omega_0 = 1.0$, goes through a second negative peak, and finally becomes zero as Ω_0 increases for a given value of Ω_b. This wave becomes stronger with the static magnetic field.

Figure C.4 relates to the components of the steady-state free-space reflected fields for the case of the switched-on magnetoplasma half-space discussed in Section C.4. In Figure C.4a, R_A corresponds to the reflection coefficient from free space to plasma. R_A is 1.0 for $\Omega_0 < \Omega_{C1}$, is zero at $\Omega_0 = 1.0$, and increases to 1.0 at $\Omega_0 = \Omega_{uh}$. It is unity up to $\Omega_0 = \Omega_{C2}$ and from there on decreases to zero. The cutoff frequencies are 0.618 and 1.618, and the upper hybrid frequency is 1.414 for $\Omega_b = 1.0$. T_{S3} refers to the transmission coefficient from plasma to free space for the first negatively propagating wave generated in the plasma. Its frequency is Ω_1. T_{S3} is zero at $\Omega_0 = 0.0$ and increases to unity as Ω_0 increases. T_{S4} corresponds to the transmission coefficient of the second negatively propagating wave having frequency Ω_2. It also starts at 0.0 at $\Omega_0 = 0.0$, but increases to 2.0 as Ω_0 increases. The relative amplitudes of the electric fields of the two waves mentioned above are shown by the curves R_{T3} and R_{T4} in Figure C.4b. R_{T3} starts at 0.13 for $\Omega_0 = 0.0$ and ultimately becomes zero. R_{T4} starts at 0.36 and reduces to zero at Ω_0. R_{T4} exhibits a negative peak when $\Omega_0 > 1.0$, but becomes zero as Ω_0 becomes large. The relative amplitudes of the corresponding free-space fields are shown by R_{B3} and R_{B4}. Both these curves start at zero and ultimately become zero as Ω_0 increases. But while R_{B3} is always positive, R_{B4} goes through a positive peak, becomes zero at $\Omega_0 = 1.0$, and goes through a negative peak afterward. R_{B3} is found to have a peak value of 0.08 at $\Omega_0 = 1.2$, $\Omega_b = 1.0$. R_{B4} has a positive peak of 0.12 at $\Omega_0 = 0.25$, $\Omega_b = 1.0$, and a negative peak of 0.03 at $\Omega_0 = 1.6$, $\Omega_b = 1.0$.

Figure C.5 compares the transient reflected electric field $A_{1R}(t)/E_0$ with the superposition of R_A, R_{B3}, and R_{B4} components. The initial peaks in the transient solution are due to the switching on of the magnetoplasma half-space. It is found that these switching transients die out quickly, within $\omega_p t = 40$ for $\Omega_b = 1.0$ and from then on there is a fair agreement between $A_{1R}(t)$ and the superposition of the three steady-state components. The low-frequency part of these curves corresponds to the R_A component. Superposed on this are the two R_B components having frequencies $|\Omega_3|$ and $|\Omega_4|$. Also, the major contribution to the reflected field is from the steady-state space boundary R_A since R_{B3} and R_{B4} are only a fraction of R_A.

When a collision frequency $\Omega_c = 0.001$ is introduced, the fields produced because of the time discontinuity are shown to decay exponentially with time. Figure C.6 shows the dependence of the decay time constants $\omega_p t_p$ of the plasma waves with Ω_0. Figure C.6a shows the variation of $\omega_p t_{p3}$ with Ω_0 for the first reflected wave of frequency Ω_1. It is found that the stronger the imposed magnetic field, the quicker the decay of this wave, for any incident wave frequency. Figure C.6b shows the decay of the second reflected wave having frequency Ω_2. This wave is absent for $\Omega_b = 0.0$. For a given Ω_0

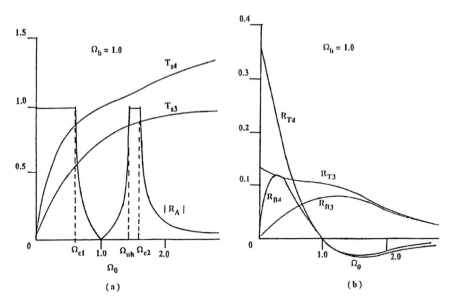

FIGURE C.4   Reflection coefficients vs. Ω_0.

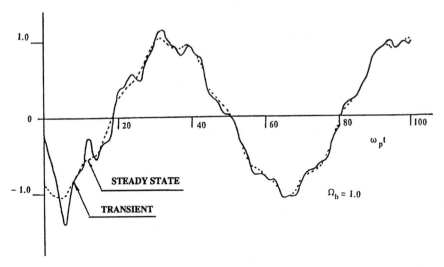

FIGURE C.5   Steady-state and transient free-space reflected electric field vs. normalized time $\omega_p t$.

and Ω_b, this wave is observed to persist longer in time than the first reflected wave. Also, as Ω_0 increases, $\omega_p t_{p2}$ (or $\omega_p t_{p4}$) attains a peak value and then finally decreases to a specific value. This final value is about 134 for $\Omega_b = 1.0$. The peak value corresponds to the incident wave frequency that gives the longest time of persistence of these waves. The peak value is about 480 at $\Omega_0 = 0.6$ for $\Omega_b = 1.0$. Also the decay of this wave is found to be faster with weaker magnetic fields.

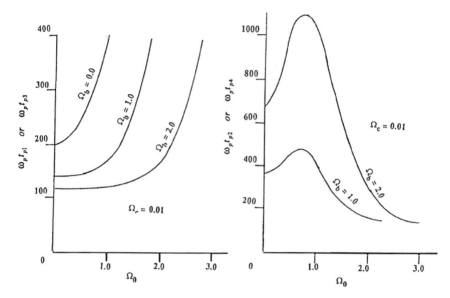

FIGURE C.6 Normalized damping time constant $\omega_p t_p$ of the plasma waves vs. Ω_0.

Figure C.7 explains the attenuation of the waves with distance as they propagate in the lossy plasma. Here the attenuation depth Z_p is normalized with the free-space wavelength corresponding to the plasma frequency, $\lambda_p (= 2\pi c/\omega_p)$. The Z_{p3}/λ_p curve of Figure C.7a is for the first reflected wave, and the Z_{p4}/λ_p curve for the second reflected wave. The dependence of Z_p/λ_p in these curves appears to follow the variation of the corresponding $\omega_p t_p$ curve. The first reflected wave is observed to attenuate over shorter distances as Ω_b is increased (Figure C.7a), while the second reflected wave (Figure C.7b) is observed to attenuate over shorter distances as Ω_b is reduced. Also, a peak attenuation depth is a characteristic of the $Z_{p4}\lambda_p$ curve. When $\Omega_b = 1.0$, the peak value of Z_{p4}/λ_p is 40 occurring at $\Omega_0 = 0.8$.

C.7 CONCLUSIONS

It is shown that the sudden creation of a magnetoplasma medium (supporting transverse propagation) splits an existing wave (incident wave) into four propagating waves having frequencies different from the incident wave frequency. Two of the newly created waves are transmitted waves and two are reflected waves. The newly created waves are attenuated and damped when the magnetoplasma is lossy. The frequencies, the damping rates, and the power carried by the new waves are governed by the magnetoplasma parameters and the incident wave parameters.

It appears that the second wave has the best chance of being observed in an experiment since it carries considerable power (Figure C.3b) and lasts a long time (Figure C.6b) for a value of $\Omega_0 \approx 0.67$. The optimum parameters for a suitable experiment may now be specified in terms of the rise time t_0 of a rapidly rising

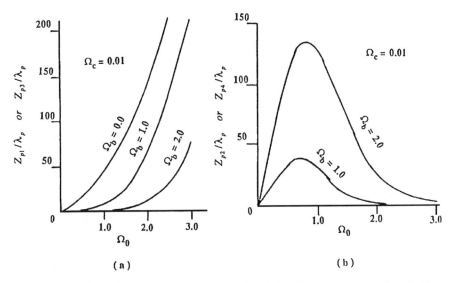

FIGURE C.7   Normalized attenuation length Z_p/λ_p of the plasma waves vs. Ω_0. The free-space wavelength λ_p corresponds to the plasma frequency.

TABLE C.1
Optimum Parameters for a Suitable Experiment

Rise Time, t_0	Incident Wave Frequency, $\omega_0/2\pi$	Plasma Frequency, $\omega_p/2\pi$	Gyrofrequency, $\omega_b/2\pi$	Collision Frequency, $\nu/2\pi$	Frequency,[a] $\omega_2/2\pi$	Time Decay Constant,[a] $t_p/2\pi$
1 μs	10 kHz	15 kHz	15 kHz	150 s^{-1}	12.45 kHz	5.6 ms
1 ns	10 MHz	15 MHz	15 MHz	150×10^3 s^{-1}	12.45 MHz	5.6 μs

[a] Values of the second wave.

particle density. By assuming that the period of the incident wave should at least be 100 times t_0 in order for the plasma to be considered as suddenly created, the incident wave frequency can be obtained as $\omega_0 = (2\pi/100t_0)$. Therefore, the plasma frequency should be $\omega_p = (3\pi/100t_0)$, and with the choice $\Omega_b = 1.0$, the gyrofrequency should be $\omega_b = (3\pi/100t_0)$. For this choice of these parameters, the frequency of the second wave, from Equation C.17b is $\omega_2 = (2.482\pi/100t_0)$, and for $\Omega_c = 0.01$ the collision frequency is $\nu = (3\pi \times 10^{-4}/t_0)$. From Figure C.6, the time decay constant of the second wave is $t_{p2} = 5623t_0$. Table C.1 shows the required parameters for two values of the rise time t_0.

REFERENCES

1. Felsen, L. B. and Whitman, G. M., Wave propagation in time-varying media, *IEEE Trans. Antennas Propag.*, AP-18, 242, 1970.
2. Fante, R. L., Transmission of electromagnetic waves into time-varying media, *IEEE Trans. Antennas Propag.*, AP-19, 417, 1971.
3. Jiang, C. L., Wave propagation and dipole radiation in a suddenly created plasma, *IEEE Trans. Antennas Propag.*, AP-23, 83, 1975.
4. Auld, B. A., Collins, J. H., and Zapp, H. R., Signal processing in a nonperiodically time-varying magnetoelastic medium, *Proc. IEEE*, 56, 258, 1968.
5. Kalluri, D. K., On reflection from a suddenly created plasma half-space: transient solution, *IEEE Trans. Plasma Sci.*, 16, 11, 1988.
6. Kalluri, D. K., Effect of switching a magnetoplasma medium on a traveling wave: longitudinal propagation, *IEEE Trans. Antennas Propag.*, AP-37, 1638, 1989.
7. Booker, H. G., *Cold Plasma Waves*, Kluwer, Hingham, MA, 1984, 349.
8. Heald, M. A. and Wharton, C. B., *Plasma Diagnostics with Microwaves,* Wiley, New York, 1965.
9. Haught, A. F., Polak, D. H., and Fader, W. J., Magnetic field confinement of laser irradiated solid particle plasmas, *Phys. Fluids*, 13, 2842, 1970.

Appendix D
Frequency Shifting Using Magnetoplasma Medium: Flash Ionization*

Dikshitulu K. Kalluri

Abstract — The main effect of switching a magnetoplasma medium is the splitting of the original wave (incident wave) into new waves whose frequencies are different from the frequency of the incident wave. The strength and the direction of the static magnetic field influence the shift ratio and the efficiency of the frequency-shifting operation. Frequency-shifting characteristics of the various waves generated in the switched magnetoplasma medium are examined.

D.I INTRODUCTION

Jiang[1] investigated the effect of switching an unbounded isotropic cold plasma medium suddenly. Since the plasma was assumed to be unbounded, the problem was modeled as a pure initial value problem. The wave propagation in such a medium is governed by the conservation of the wave number across the time discontinuity and consequent change in the wave frequency. Rapid creation of the medium can be approximated as a sudden switching of the medium. The main effect of switching the medium is the splitting of the original wave (henceforth called incident wave, in the sense of incidence on a time discontinuity in the properties of the medium) into new waves whose frequencies are different from the incident wave. Recent experimental work[2-4] and computer simulation[5] demonstrated the frequency upshifting of microwave radiation by rapid plasma creation.

In any practical situation, the plasma is bounded and the problem can no longer be modeled as a pure initial value problem. Kalluri[6] dealt with this aspect by considering the reflection of a traveling wave when an isotropic cold plasma is switched on only over $z > 0$ half-space. Kalluri has shown that the reflected field in free space comprises two components, A and B. The A component is due to reflection at the spatial discontinuity at $z = 0$ formed at $t = 0$. Its frequency is the same as that of the incident wave. The temporal discontinuity gives rise to two additional waves in the plasma. These are called B waves. One of them propagates in the negative

---

* © 1993 IEEE. Reprinted with permission from Kalluri, D. K., *IEEE Transactions on Plasma Science*, 21, 78–81, 1993.

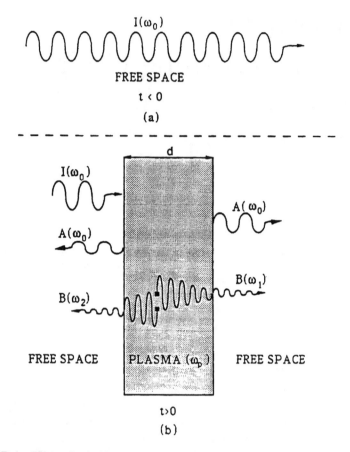

FIGURE D.1   Effect of switching an isotropic plasma slab. Reflected and transmitted waves are sketched. A waves have the same frequency as the incident wave frequency, but B waves have a new frequency $\omega_1 = (\omega_0^2 + \omega_p^2)^{1/2} = |\omega_2|$.

z-direction, and it is this wave that undergoes partial transmission into free space. The frequency of the B component is different from that of the incident wave.

Kalluri and Goteti[7] recently brought the solved problem closer to the practical situation by considering the switched plasma medium as a slab of width d (Figure D.1). A qualitative picture of the reflected and transmitted waves is given in the figure. If the incident wave frequency ω_0 is less than the plasma wave frequency ω_p, i.e., ω_0 is in the stop band of the medium, the transmitted wave will be a B wave of the new[1] frequency $\omega_1 = \sqrt{\omega_0^2 + \omega_p^2}$.

The solution of the slab problem was complex, but it would be more complex if the posed problem was to be brought even closer to the practical situation, i.e., the problem of switching a space-varying and time-varying plasma of finite extent. However, since the B waves are generated in the plasma, their basic nature in terms of the frequency shift, power, and energy in them and how they are damped can all be studied qualitatively by considering the switching of the unbounded medium.

D.2 WAVE PROPAGATION IN A SWITCHED, UNBOUNDED MAGNETOPLASMA MEDIUM

It is known[8] that the presence of a static magnetic field will influence the frequency shift, etc. Rapid creation of a plasma medium in the presence of a static magnetic field can be approximated as the sudden switching of a magnetoplasma medium. An idealized mathematical description of the problem is given next. A uniform plane wave is traveling in free space along the z-direction, and it is assumed that, at $t =$ 0^-, the wave occupies the entire space. A static magnetic field of strength $\mathbf{B}_0$ is assumed to be present throughout. At $t = 0$, the entire space is suddenly ionized, thus converting the medium from free space to a magnetoplasma medium. The incident wave frequency is assumed to be high enough that the effects of ion motion may be neglected.

The strength and the direction of the static magnetic field affect the number of new waves created, their frequencies, and the power density of these waves. Some of the new waves are transmitted waves (waves propagating in the same direction as the incident wave), and some are reflected waves (waves propagating in the opposite direction to that of the incident wave). Reflected waves tend to have less power density.

A physical interpretation of the waves may be given in the following way. The electric and magnetic fields of the incident wave and the static magnetic field accelerate the electrons in the newly created magnetoplasma, which in turn radiate new waves. The frequencies of the new waves and their fields can be obtained by adding contributions from the many electrons whose positions and motions are correlated by the collective effects supported by the magnetoplasma medium. Such a detailed calculation of the radiated fields seems to be quite involved. A simple, but less accurate, description of the plasma effect is obtained by modeling the magnetoplasma as a dielectric medium whose refractive index is computed through magnetoionic theory.[9] The frequencies of the new waves are constrained by the requirements that the wave number (k_0) is conserved over the time discontinuity and the refractive index n is that applicable to the type of wave propagation in the magnetoplasma. This gives a conservation[10] law $k_0 c = \omega_0 = n(\omega)\omega$ from which ω can be determined. Solution of the associated electromagnetic initial value problem gives the electric and magnetic fields of the new waves. By using this approach, the general aspects of wave propagation in a switched magnetoplasma medium were discussed earlier[8,11,12] for the cases of L, R, and X incidence.[9]

D.3 L AND R WAVES

L incidence is an abbreviation used to describe the case where the incident wave has left-hand circular polarization and the static magnetic field is in the z-direction (Figure D.2a). Three new waves with left-hand circular polarization labeled as $L1$, $L2$, and $L3$ are generated by the medium switching. $L1$ is a transmitted wave. R incidence refers to the case where the incident wave has right-hand circular polarization and the static magnetic field is in the z-direction (Figure D.2b). The medium switching, in this case, creates three R waves labeled $R1$, $R2$, and $R3$. $R1$ and $R3$

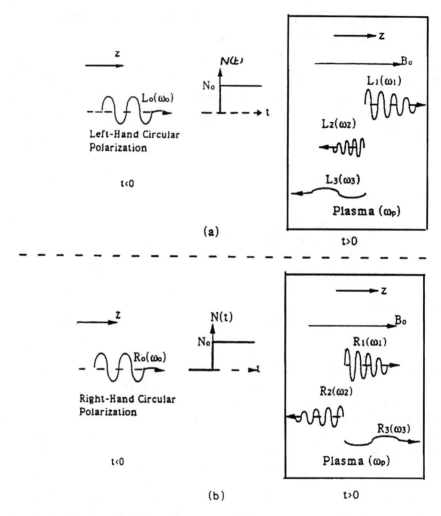

FIGURE D.2 Effect of switching an unbounded magnetoplasma medium. Sketches of the B waves generated in the plasma are given for (a) L-incidence (b) R-incidence.

are transmitted waves. The frequencies of the waves are the roots of the characteristic equation.[9]

$$\omega^3 \mp \omega_b \omega^2 - \left(\omega_0^2 + \omega_p^2\right)\omega \pm \omega_0^2 \omega_b = \left(\omega - \omega_1\right)\left(\omega - \omega_2\right)\left(\omega - \omega_3\right) = 0, \quad \text{(D.1)}$$

where the lower sign is for the L waves and the upper sign is for the R waves. The electric and the magnetic fields of the nth wave are given in Equation D.2 and D.3.

$$\frac{E_n}{E_0} = \frac{\left(\omega_n \mp \omega_b\right)\left(\omega_n + \omega_0\right)}{\displaystyle\prod_{m=1,m\neq n}^{3} \left(\omega_n - \omega_m\right)}, \quad n = 1, 2, 3, \quad \text{(D.2)}$$

$$\frac{H_n}{H_0} = \frac{\left(\omega_n \mp \omega_b\right)\left(\omega_n + \omega_0\right) - \omega_p^2}{\displaystyle\prod_{m=1,m\neq n}^{3}\left(\omega_n - \omega_m\right)}, \quad n = 1, 2, 3. \tag{D.3}$$

In the above, ω_b is the electron gyrofrequency and the incident wave variables are denoted by zero subscript. A negative value for $\omega_n{}^8$ indicates that the wave is a reflected wave; i.e., it propagates in a direction opposite to that of the source (incident) wave. If the source wave is right going, then the reflected wave will be left going.

The following approximations can be made when $\omega_b \ll \omega_0$:

$$\omega_1 = \sqrt{\omega_0^2 + \omega_p^2} \pm \frac{\omega_b}{2}\frac{\omega_p^2}{\omega_0^2 + \omega_p^2}, \quad \omega_b \ll \omega_0, \tag{D.4a}$$

$$\omega_2 = -\sqrt{\omega_0^2 + \omega_p^2} \pm \frac{\omega_b}{2}\frac{\omega_p^2}{\omega_0^2 + \omega_p^2}, \quad \omega_b \ll \omega_0, \tag{D.4b}$$

$$\omega_3 = \pm\omega_b \frac{\omega_0^2}{\omega_0^2 + \omega_p^2}, \quad \omega_b \ll \omega_0, \tag{D.4c}$$

$$\frac{E_3}{E_0} = \pm\omega_b \frac{\omega_0 \omega_p^2}{\left(\omega_0^2 + \omega_p^2\right)^2}, \quad \omega_b \ll \omega_0, \tag{D.5a}$$

$$\frac{H_3}{H_0} = \frac{E_3}{E_0} + \frac{\omega_p^2}{\left(\omega_0^2 + \omega_p^2\right)}, \quad \omega_b \ll \omega_0, \tag{D.5b}$$

$$\frac{S_3}{S_0} = \frac{E_3 H_3}{E_0 H_0} = \pm\frac{\omega_0 \omega_b \omega_p^4}{\left(\omega_0^2 + \omega_p^2\right)^3}, \quad \omega_b \ll \omega_0. \tag{D.5c}$$

The first terms on the right side of Equations D.4a and b are the frequencies for the isotropic case. The approximations are quite good (error less than 5%) for ω_b/ω_0 as high as 0.5.

D.4   X WAVES

X incidence refers to the case where the electric field of the incident wave is in the x-direction, and the static magnetic field is in the y-direction. The medium switching in this case creates four extraordinary waves labeled X1, X2, X3, and X4. X1 and X2 are transmitted waves. The frequencies of these waves are

$$\omega_1 = \sqrt{A + \sqrt{A^2 - B}}, \tag{D.6a}$$

$$\omega_2 = \sqrt{A - \sqrt{A^2 - B}}, \tag{D.6b}$$

$$\omega_3 = -\omega_1, \tag{D.6c}$$

$$\omega_4 = -\omega_2, \tag{D.6d}$$

where

$$A = \left(\omega_0^2 + \omega_b^2 + 2\omega_p^2\right)/2, \tag{D.6e}$$

$$B = \omega_p^4 + \omega_0^2\left(\omega_p^2 + \omega_b^2\right). \tag{D.6f}$$

The fields for this case are given in Reference 11. The following approximations can be made when $\omega_b \ll \omega_0$:

$$\omega_1 = \sqrt{\omega_0^2 + \omega_p^2} + \frac{\omega_b^2}{2}\frac{\omega_p^2}{\omega_0^2\sqrt{\omega_0^2 + \omega_p^2}}, \qquad \omega_b \ll \omega_0, \tag{D.7a}$$

$$\omega_2 = \sqrt{\omega_b^2 + \omega_p^2} - \frac{\omega_b^2}{2}\frac{\omega_p^2}{\omega_0^2\sqrt{\omega_b^2 + \omega_p^2}}, \qquad \omega_b \ll \omega_0. \tag{D.7b}$$

The approximations are quite good (error less than 5%) for ω_b/ω_0 as high as 0.5. A further approximation for ω_2 can be made if ω_b is also much less than ω_p:

$$\omega_2 = \omega_p + \frac{1}{2}\frac{\omega_b^2}{\omega_0^2}\frac{\omega_0^2 - \omega_p^2}{\omega_p}, \qquad \omega_b \ll \omega_0 \text{ and } \omega_b \ll \omega_p. \tag{D.7c}$$

The first terms on the right side of Equation D.7a and c are the frequencies for the isotropic case.

D.5   FREQUENCY-SHIFTING CHARACTERISTICS OF VARIOUS WAVES

The shift ratio and the efficiency of the frequency-shifting operation may be controlled by the strength and the direction of the static magnetic field. It is this aspect that is emphasized in this appendix. The results are presented by normalizing all frequency variables with respect to the incident wave frequency (source frequency)

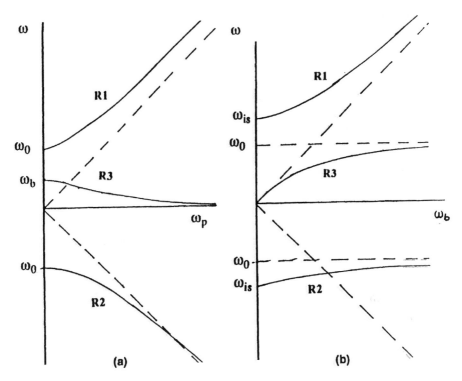

FIGURE D.3 Frequency shifting of R waves. Sketch of (a) ω vs. ω_p, (b) ω vs. ω_b. $\omega_{is} = (\omega_0^2 + \omega_p^2)^{1/2}$. Here ω_0 is the incident wave frequency, ω_p is the plasma frequency, and ω_b is the electron gyrofrequency.

ω_0. This normalization is achieved by taking $\omega_0 = 1$ in numerical computations. Here, $\omega_b = qB_0/m$ is the electron gyrofrequency, ω_0 is the frequency (angular) of the source (incident) wave, ω_n is the frequency of the nth wave generated by the medium switching, and the other symbols have the usual meaning.[8]

For R waves, the curves of ω vs. ω_p and ω vs. ω_b are sketched in Figure D.3a and b, respectively. The sketches for the $L1$, $L2$, and $L3$ waves are mirror images (with respect to the horizontal axis) of those of $R2$, $R1$, and $R3$ waves, respectively. Figure D.4 shows the sketches for the X waves.

In Figure D.5 the results are presented for the $R1$ wave: the values on the vertical axis give the frequency-shift ratio since the frequency variables are normalized with respect to ω_0. This is an upshifted wave, and both ω_p and ω_b improve the shift ratio. From this figure it appears that, by a suitable choice of ω_p and ω_b, it is possible to obtain any desired large frequency shift. However, the wave generated may have weak fields associated with it and the power density S_1 may be low. This point is illustrated in Table D.1 by considering two sets of values for the parameters (ω_p, ω_b). For the set (0.5, 0.5) the shift ratio is 1.2, but the power density ratio S_1/S_0 is 57.3%, whereas for the set (2.0, 2.0), the shift ratio is 3.33, but the power density ratio is only 7%. Similar remarks apply to other waves. A detailed report on this aspect can be found elsewhere (see Chapter 7).

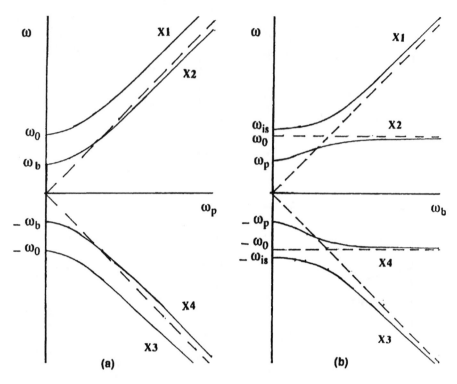

FIGURE D.4 Frequency shifting of the X wave. Sketch of (a) ω vs. ω_p, (b) ω vs. ω_b. The meaning of the symbols is given in Figure D.3.

Figure D.6 shows the magnitude of the frequency of the $R2$ wave. This is a reflected wave. The $L1$ wave is a transmitted wave with the same frequency characteristics. This is an upshifted wave, and the shift ratio increases with ω_p, but decreases with ω_b.

The $R3$ wave (Figure D.7) is a transmitted wave that is downshifted. The shift ratio decreases with ω_p and increases with ω_b. When $\omega_b = 0$, ω_{R3} becomes zero. The electric field E_3 becomes zero, and the magnetic field degenerates to the wiggler magnetic field.[8]

Figures D.8 and D.9 describe the characteristics of X waves. The $X1$ wave is a transmitted wave, and the shift ratio increases with ω_p and ω_b. The $X2$ wave is downshifted for $\omega_p < \omega_0$ and upshifted for $\omega_p > \omega_0$. The shift ratio increases with ω_b for $\omega_p < \omega_0$ and decreases with ω_b for $\omega_p > \omega_0$. Attention is drawn to the curve with the parameter $\omega_b = 0$ in Figure D.9a. For this case, $\omega_{X2} = \omega_p$. However, it can be shown[11] that all the fields associated with this wave are zero. From Figure D.9b, the shift ratio is unity when $\omega_p = \omega_0$ for all values of ω_b, but the power density increases[11] with ω_b.

ACKNOWLEDGMENT

The author thanks W. Mori and C. Joshi for their useful comments.

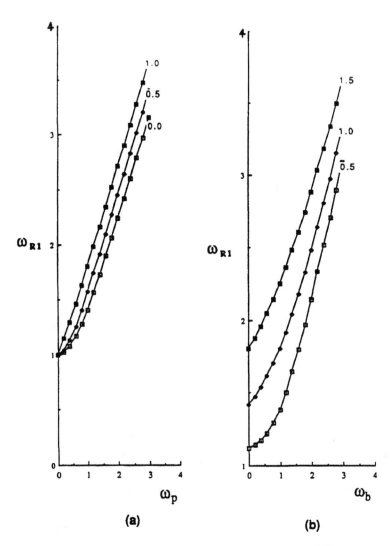

FIGURE D.5 Frequency shifting of the $R1$ wave. The frequency variables are normalized with respect to incident wave frequency by taking $\omega_0 = 1$ in numerical computations. The numbers on the curves are values of the parameters (a) ω_b, (b) ω_p.

TABLE D.1
$R1$ Wave Shift Ratio and Power Density
for Two Sets of (ω_p, ω_b)

ω_0	ω_p	ω_b	E_1/E_0	H_1/H_0	S_1/S_0	ω_{R1}/ω_0
1	0.5	0.5	0.83	0.69	57.3%	1.20
1	2.0	2.0	0.39	0.18	7.0%	3.33

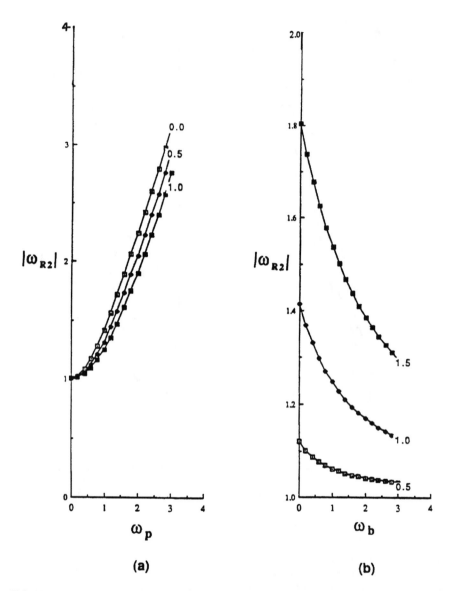

FIGURE D.6 Frequency shifting of the $R2$ wave. $\omega_0 = 1$. This is a reflected wave and the vertical axis gives the magnitude of ω_{R2}. The numbers on the curves are values of the parameters (a) ω_b, (b) ω_p.

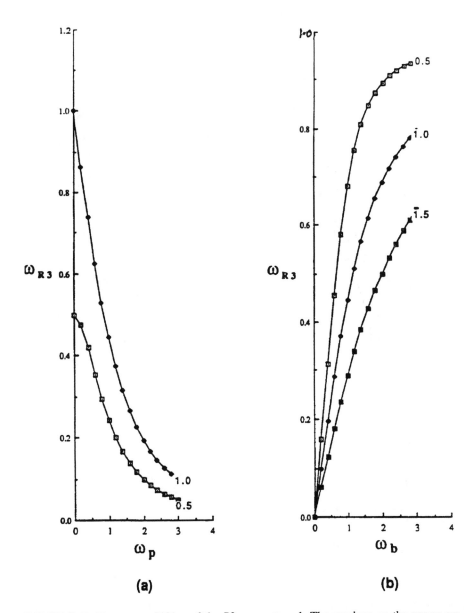

FIGURE D.7 Frequency shifting of the $R3$ wave. $\omega_0 = 1$. The numbers on the curves are values of the parameters (a) ω_b, (b) ω_p.

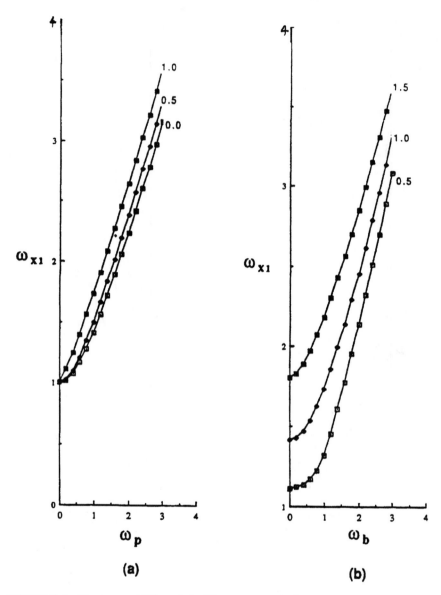

FIGURE D.8 Frequency shifting of the $X1$ wave. $\omega_0 = 1$. The numbers on the curves are values of the parameters (a) ω_b, (b) ω_p.

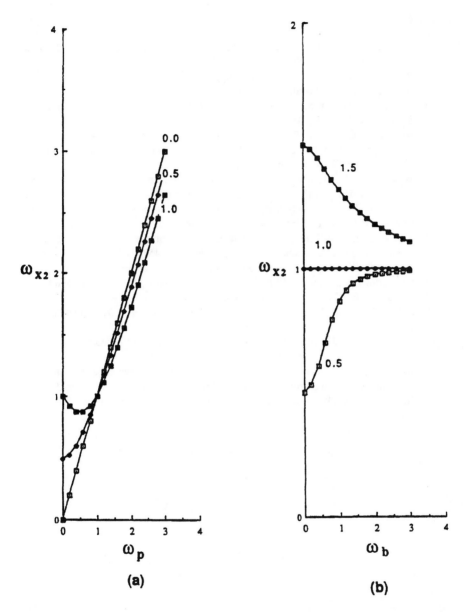

FIGURE D.9 Frequency shifting of the $X2$ wave. $\omega_0 = 1$. The numbers on the curves are values of the parameters (a) ω_b, (b) ω_p.

REFERENCES

1. Jiang, C. L., Wave propagation and dipole radiation in a suddenly created plasma, *IEEE Trans. Antennas Propag.*, AP-23, 83, 1975.
2. Rader, M., Dyer, F., Matas, A., and Alexeff, I., in *Conf. Rec. Abstr., IEEE Int. Conf. Plasma Science*, Oakland, CA, 1990, 171.
3. Kuo, S. P., Zhang, Y. S., and Ren, A. Q., in *Conf. Rec. Abstr., IEEE Int. Conf. Plasma Science*, Oakland, CA, 1990, 171.
4. Joshi, C. J., Clayton, C. E., Marsh, K., Hopkins, D. B., Sessler, A., and Whittum, D., Demonstration of the frequency upshifting of microwave radiation by rapid plasma creation, *IEEE Trans. Plasma Sci.*, 18, 814, 1990.
5. Wilks, S. C., Dawson, J. M., and Mori, W. B., Frequency up-conversion of electromagnetic radiation with use of an overdense plasma, *Phys. Rev. Lett.*, 61, 337, 1988.
6. Kalluri, D. K., On reflection from a suddenly created plasma half-space: transient solution, *IEEE Trans. Plasma Sci.*, 16, 11, 1988.
7. Kalluri, D. K. and Goteti, V. R., Frequency shifting of electromagnetic radiation by sudden creation of a plasma slab, *J. Appl. Phys.*, 72, 4575, 1992.
8. Kalluri, D. K., Effect of switching a magnetoplasma medium on a traveling wave: longitudinal propagation, *IEEE Trans. Antennas Propag.*, AP-37, 1638, 1989.
9. Booker, H. G., *Cold Plasma Waves*, Kluwer, Hingham, MA, 1984.
10. Kalluri, D. K., in *Conf. Rec. Abstr., IEEE Int. Conf. Plasma Science*, Oakland, CA, 1990, 129.
11. Goteti, V. R. and Kalluri, D. K., Wave propagation in a switched magnetoplasma medium: transverse propagation, *Radio Sci.*, 25, 61, 1990.
12. Kalluri, D. K. and Goteti, V. R., Damping rates of waves in a switched magnetoplasma medium: longitudinal propagation, *IEEE Trans. Plasma Sci.*, 18, 797, 1990.
13. Heald, M. A. and Wharton, C. B., *Plasma Diagnostics with Microwaves*, Wiley, New York, 1965, 36.

Appendix E
Frequency Upshifting with Power Intensification of a Whistler Wave by a Collapsing Plasma Medium*

Dikshitulu K. Kalluri

Abstract — A source wave is assumed to be present in an unbounded magnetoplasma medium. The parameters are such that the refractive index is greater than 1. A wave propagating in the *whistler mode* is an example of such a wave. By collapsing the ionization, it is possible to obtain a frequency-upshifted wave with power intensification. Two limiting cases of (1) sudden collapse and (2) slow decay are considered. In either case, it is shown that the final upshifted frequency is the source frequency multiplied by the refractive index. When the source frequency is much less than electron gyrofrequency, the refractive index n_w is quite large and the electric field is intensified by a factor of $n_w/2$ for the case of sudden collapse and by a factor of $n_w/\sqrt{2}$ for the case of slow decay. The corresponding intensification factors for the power density are $n_w/4$ and $n_w/2$. A physical explanation of the results, based on energy balance, is offered.

E.1 INTRODUCTION

Frequency upshifting using time-varying *magnetoplasmas* has been investigated[1-6] recently. In all these cases the electron density profile was a monotonically increasing function of time. A fast profile was modeled as sudden creation of a plasma medium. The slow profile problem was solved using the Wentzel–Kramers–Brillouin (WKB) method. The calculations showed that the power density in the frequency-shifted waves was less than the power density of the source wave.

It is known[7] that across a temporal discontinuity in the properties of the medium, while the energy is conserved, the sum of the power densities of the new waves may be less than or greater than the power density of the source wave. The case of "greater power density in the new wave" with upshifted frequency seems to occur under the

---

* © 1996 American Institute of Physics. Reprinted with permission from D. K. Kalluri, *Journal of Applied Physics*, 79, 3895–3899, 1996.

following circumstances: A right circularly polarized electromagnetic wave (R wave) is propagating in the whistler mode in an unbounded magnetoplasma medium. At $t = 0$, the electron density starts decreasing, and, for $t > 0$, the electron density profile is a monotonically decreasing function of time.

E.2   SUDDEN COLLAPSE

An R wave is assumed to be propagating in the positive z-direction in a spatially unbounded magnetoplasma medium. The electric and magnetic fields of this wave, named as the source wave, are

$$\bar{e}_i(z,t) = (\hat{x} - j\hat{y})E_0 \exp\left[j(\omega_0 t - \beta_0 z)\right],\qquad\text{(E.1a)}$$

$$\bar{h}_i(z,t) = (j\hat{x} + \hat{y})H_0 \exp\left[j(\omega_0 t - \beta_0 z)\right],\qquad\text{(E.1b)}$$

where ω_0 is the frequency of the incident wave and

$$\beta_0 = \omega_0 n_0 / c,\qquad\text{(E.2a)}$$

$$n_0\sqrt{\left(1 - \frac{\omega_{p0}^2/\omega_0^2}{1 - \omega_b/\omega}\right)},\qquad\text{(E.2b)}$$

Here, c is the velocity of light in free space, ω_{p0} is the plasma frequency, ω_b is the electron gyrofrequency, and n_0 is the refractive index of the magnetoplasma medium existing prior to $t = 0$. Assume that at $t = 0$ the electron density undergoes a step change and the corresponding new plasma frequency for $t > 0$ is ω_{p1}. The new refractive index n_1 is given by

$$n_1\sqrt{\left(1 - \frac{\omega_{p1}^2/\omega^2}{1 - \omega_b/\omega}\right)},\qquad\text{(E.3)}$$

where ω is the frequency of the new waves created by the switching.

E.2.1   FREQUENCIES OF THE NEW WAVES

Because the wave number is conserved across the temporal discontinuity,[7,8]

$$\omega^2 n_1^2 = \omega_0^2 n_0^2.\qquad\text{(E.4)}$$

Equation E.4 is a cubic in ω. The frequencies of the three new waves generated by the switching action are denoted ω_1, ω_2, and ω_3. A negative value for ω indicates a reflected wave traveling in a direction opposite to that of the source wave.

A fast collapse of the plasma can be idealized as a sudden collapse, and the frequencies of the new waves in this case are obtained by substituting $\omega_{p1} = 0$ in Equation E.4, giving

$$\omega_1 = n_0\omega_0, \omega_2 = -n_0\omega_0, \text{ and } \omega_3 = \omega_b. \tag{E.5}$$

E.2.2 FIELDS OF THE NEW WAVES

The fields of the new waves can be determined by imposing the initial conditions that the E and H fields are continuous over the temporal discontinuity:

$$E_1 + E_2 + E_3 = E_0, \tag{E.6a}$$

$$H_1 + H_2 + H_3 = H_0. \tag{E.6b}$$

The third equation, which should come from the initial condition of continuity of the velocity field, is not obvious since it is assumed that the plasma has suddenly collapsed. The author has recently completed a study wherein a gradual collapse of the electron density, i.e., that $\omega_p(t)$ is a monotonically decreasing function of time, was assumed. The study clearly showed that the frequency of the third wave tends to ω_b and the fields E_3 and H_3 tend to 0 as t goes to infinity. The third wave, which is supported by only a magnetoplasma medium, vanishes in free space. Thus, $E_3 = H_3 = 0$ in Equation E.6. Noting that $H_1 = E_1/120\pi$, $H_2 = -E_2/120\pi$, and $H_0 = E_0n_0/120\pi$, the following is obtained:

$$E_1/E_0 = (1+n_0)/2, \qquad E_2/E_0 = (1-n_0)/2, \tag{E.7a}$$

$$H_1/H_0 = (1+n_0)/2n_0, \quad H_2/H_0 = -(1-n_0)/2n_0, \tag{E.7b}$$

$$S_1/S_0 = (1+n_0)^2/4n_0, \quad S_2/S_0 = -(1-n_0)^2/4n_0. \tag{E.7c}$$

Here, the symbol S stands for the power density in the wave, and the subscript indicates the appropriate wave.

E.2.3 CHOICE OF THE PARAMETERS

In the whistler mode,[9,10] i.e., $\omega_0 < \omega_b$, n_0 is greater than 1, and therefore $\omega_1/\omega_0 > 1$, $E_1/E_0 > 1$, and $S_1/S_0 > 1$. Thus, there is a frequency-upshifted wave with power intensification.

The frequency upshift is large near cyclotron resonance, i.e., when ω_0 is approximately equal to ω_b, but in this neighborhood the effect of collisions cannot be ignored. There is a trade-off in the choices of ω_p/ω_0 and ω_b/ω_0. For a low-loss magnetoplasma of collision frequency ν, the refractive index is complex and given by[10]

$$\tilde{n}_0 = n_0 - j\chi_0 = \sqrt{1 - \frac{\omega_{p0}^2/\omega_0^2}{1 - \omega_b/\omega_0 - j\nu/\omega_0}}. \tag{E.8}$$

As ω_b/ω_0 approaches 1, n_0 increases but the attenuation constant $\alpha = \chi_0(\omega_0/c)$ also increases, and therefore it will be difficult to maintain the source wave before $t = 0$. For a sample calculation, take $\nu/\omega_0 = 0.01$ and $\omega_p/\omega_0 = 0.8$; for $\omega_b/\omega_0 = 1.2$, $n_0 = 2.05$ and $\alpha = 0.039$ (ω_0/c); for $\omega_b/\omega_0 = 1.1$, $n_0 = 2.71$, but $\alpha = 0.117$ (ω_0/c). For a given ω_b/ω_0, as ω_p/ω_0 increases, n_0 increases, but the attenuation constant also increases. Again, for a sample calculation, take $\nu/\omega_0 = 0.01$ and $\omega_b/\omega_0 = 1.2$; for $\omega_p/\omega_0 = 0.5$, $n_0 = 1.5$ and $\alpha = 0.021$(ω_0/c); for $\omega_p/\omega_0 = 0.8$, $n_0 = 2.05$ but $\alpha = 0.039$(ω_0/c).

E.2.4 ECKERSLEY APPROXIMATION

The refractive index is also large when $\omega_0 \ll \omega_b \sim \omega_{p0}$ and from Equation E.2b

$$n_0 \approx n_w = \omega_{p0}/(\omega_0\omega_b)^{1/2}, \quad \omega_0 \ll \omega_b \sim \omega_{p0}, \tag{E.9a}$$

and from Equation E.7a and c

$$E_1/E_0 \approx n_w/2, \quad \omega_0 \ll \omega_b \sim \omega_{p0}, \tag{E.9b}$$

$$H_1/H_0 \approx 1/2, \quad \omega_0 \ll \omega_b \sim \omega_{p0}, \tag{E.9c}$$

$$S_1/S_0 \approx n_w/4, \quad \omega_0 \ll \omega_b \sim \omega_{p0}. \tag{E.9d}$$

In this approximation, called the Eckersley approximation,[9] the phase and the group velocities are proportional to the square root of the signal frequency. Eckersley used this approximation to explain the phenomenon of whistlers. Whistlers are naturally occurring electromagnetic waves (due to lightning) that have frequencies in the audio band. They propagate from one hemisphere to the other through the plasmasphere along Earth's magnetic field. Since the group velocity of these waves increases with frequency, the low frequencies arrive later and give rise to the descending tone.

A physical explanation for the power intensification can be given based on energy consideration. The energy in the whistler mode, under the Eckersley approximation, is predominantly the magnetic energy due to plasma current, and its density w_m is given by

$$w_m = \tfrac{1}{2}\varepsilon_0 E_0^2 n_2^2. \tag{E.10}$$

After the plasma collapses, the plasma current collapses and the magnetic energy due to plasma current is converted into wave electric and magnetic energy giving rise to the frequency-upshifted waves with enhanced electric field and power density. The energy balance equation can be written as

$$\tfrac{1}{2}\varepsilon_0 E_1^2 + \tfrac{1}{2}\mu_0 H_1^2 + \tfrac{1}{2}\varepsilon_0 E_2^2 + \tfrac{1}{2}\mu_0 H_2^2 = 2\varepsilon_0 E_1^2 = \tfrac{1}{2}\varepsilon_0 E_0^2 n_w^2. \tag{E.11}$$

The intermediate step in Equation E.11 is explained as follows: the electric and magnetic field energy densities are equal for each of the waves. Moreover, the amplitudes of the electric fields E_1 and E_2 have approximately the same magnitude. Thus, the result $E_1/E_0 = n_w/2$ from the energy balance is obtained, in agreement with Equation E.9b.

The power intensification calculation can be made from a different viewpoint if the following least-efficient way of establishing the source wave in the magneto-plasma medium is also considered, by posing the following idealized problem. For $t < 0$, let the half-space $z < 0$ be free space in which an R wave of frequency ω_0 and fields E_{-1}, H_{-1} is propagating along the direction of the z-axis. The half-space $z > 0$ is a magnetoplasma of plasma frequency ω_{p0} and electron gyrofrequency ω_b. At $t = 0$, let the plasma suddenly collapse, giving rise to the frequency-upshifted wave. It is easily shown that $E_0/E_{-1} = 2/(1 + n_0)$, and from Equation E.7a that $E_1/E_{-1} = 1$. Similarly, $H_1/H_{-1} = 1$.

In the above calculation losses were neglected, but the point can be made that even in the less-efficient case there is a frequency transformation without a power reduction.

The key to frequency upshifting using the collapse of the magnetoplasma is the establishment of the source wave propagation in the pass band of the magnetoplasma with the refractive index $n_0 > 1$. Thus, one can use an X wave[10] rather than the R wave in the frequency range $\omega_{p0} < \omega_0 < \omega_{uh}$, where ω_{uh} is the upper hybrid frequency. The X wave is the extraordinary mode of transverse propagation in a magnetized plasma. The direction of phase propagation of this wave is perpendicular to the direction of the imposed magnetic field. The electrical field is elliptically polarized in a plane that contains the direction of propagation but is perpendicular to the imposed magnetic field. The wave magnetic field is parallel to the imposed magnetic field.

E.2.5 EXPERIMENTAL FEASIBILITY

A sudden collapse of the plasma on a half-space is an idealization of the problem and is difficult to achieve experimentally. If a slab of finite width[11] instead of a half-space is considered, the new wave will be an intensified electromagnetic pulse at the upshifted frequency. On the other hand, an experimental technique analogous to the ionization front[6,12,13] can be used. In this case, a deionization (recombination) front must be used, i.e., a process of deionization[13] that moves with a deionization front velocity V_0 close to c is created. Work is in progress in this regard.

E.3   SLOW DECAY

A WKB type of solution can be obtained for the case of slow decay; i.e., $\omega_p^2(t)$ is a slowly decaying function of time. The technique is similar to that used in References 3, 14, and 15, and, therefore, much of the mathematical detail is omitted here. The frequencies of the three waves created by the switching action are the roots of the cubic:

$$\omega^3 - \omega^2\omega_b - \left[\omega_0^2 n_0^2 + \omega_p^2(t)\right]\omega + \omega_0^2\omega_b n_0^2 = 0. \tag{E.12}$$

This cubic has two positive real roots and one negative real root. The amplitudes of the fields associated with these waves are obtained by solving the initial value problem. The continuity of the electric and magnetic fields provides two initial conditions. The third initial condition is the continuity of the velocity field at $t = 0$. The decay of the plasma with time is assumed to take place in a process of sudden capture of a number of free electrons as time passes. The velocity of the remaining free electrons is assumed to be unaffected.

The amplitudes of two of the waves are of the order of the time derivative of the $\omega_p^2(t)$ curve and, thus, negligible. The amplitude of the wave whose initial frequency is ω_0 is significant, and this wave will be called the modified source wave. Its fields $e_1(z,t)$ and $h_1(z,t)$ can be expressed as

$$e_1(z,t) = E_1(t)\exp\left[j\left(\int_0^t \omega_1(t)dt - \beta_0 z \right)\right] \tag{E.13a}$$

and

$$h_1(z,t) = H_1(t)\exp\left[j\left(\int_0^t \omega_1(t)dt - \beta_0 z \right)\right], \tag{E.13b}$$

where

$$\frac{E_1(t)}{E_0} = \frac{\omega_b - \omega_1(t)}{\omega_b - \omega_0}\frac{\omega_1(t)}{\omega_0}\sqrt{\frac{\left(\omega_b\omega_0^2 n_0^2 + \omega_b\omega_0^2 - 2\omega_0^3\right)}{\left(\omega_b\omega_0^2 n_0^2 + \omega_b\omega_1^2 - 2\omega_1^3\right)}}, \tag{E.14a}$$

$$\frac{H_1(t)}{H_0} = \frac{E_1(t)}{E_0}\frac{n(t)}{n_0}, \tag{E.14b}$$

$$n(t) = \sqrt{\left(1 - \frac{\omega_p^2(t)/\omega_1^2(t)}{1 - \omega_b/\omega_1(t)}\right)}, \tag{E.14c}$$

and

$$\frac{S_1(t)}{S_0} = \frac{E_1(t)}{E_0} \frac{H_1(t)}{H_0}. \tag{E.15}$$

After the plasma completely collapses, i.e., as $t \to \infty$,

$$\omega_1(t \to \infty) = n_0 \omega_0, \tag{E.16a}$$

$$\frac{E_1(t \to \infty)}{E_0} = \frac{\omega_b - n_0 \omega_0}{\omega_b - \omega_0} \sqrt{\frac{\omega_b(1 + n_0^2) - 2\omega_0}{(2\omega_b - 2n_0\omega_0)}}. \tag{E.16b}$$

E.3.1 ECKERSLEY APPROXIMATION

Again, when $\omega_0 \ll \omega_b \sim \omega_{p0}$, $n_0 \approx n_w$ and n_w is given by Equation E.9a. The final values for the frequency and the electric field are

$$\omega_1(t \to \infty) \approx n_w \omega_0, \tag{E.17a}$$

$$\frac{E_1(t \to \infty)}{E_0} \approx \frac{n_w}{\sqrt{2}} \left(1 - \frac{\omega_{p0}}{2\omega_b} \sqrt{\frac{\omega_0}{\omega_b}}\right) \approx \frac{n_w}{\sqrt{2}}, \tag{E.17b}$$

$$\frac{H_1(t \to \infty)}{H_0} \approx \frac{1}{\sqrt{2}} \left(1 - \frac{\omega_{p0}}{2\omega_b} \sqrt{\frac{\omega_0}{\omega_b}}\right) \approx \frac{1}{\sqrt{2}}, \tag{E.17c}$$

$$\frac{S_1(t \to \infty)}{S_0} \approx \frac{n_w}{2} \left(1 - \frac{\omega_{p0}}{\omega_b} \sqrt{\frac{\omega_0}{\omega_b}}\right) \approx \frac{n_w}{2}. \tag{E.17d}$$

The middle set of equations in Equation E.17 is the first set of approximations and the last set of approximations in Equation E.17 is obtained by approximating the middle set further.

A physical explanation is obtained from the energy balance equation:

$$\tfrac{1}{2}\varepsilon_0 E_1^2 + \tfrac{1}{2}\mu_0 H_1^2 = \varepsilon_0 E_1^2 = \tfrac{1}{2}\varepsilon_0 E_0^2 n_w^2. \tag{E.18}$$

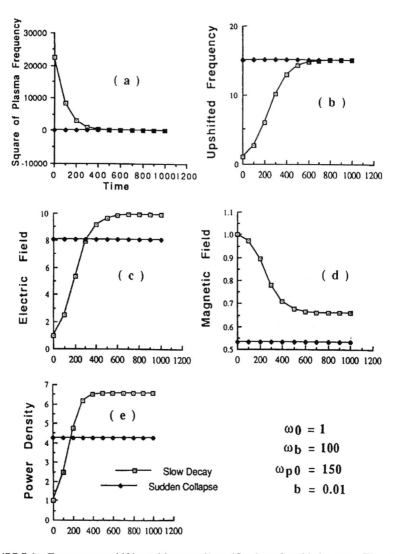

FIGURE E.1 Frequency upshifting with power intensification of a whistler wave. The values on the vertical axes are normalized with respect to source wave quantities. The source frequency ω_0 is taken as 1. The horizontal timescale is normalized with respect to the period of the source wave. The parameters are such that Eckersley approximation is applicable.

In the slow decay case under discussion, the fields of the second and third waves are insignificant; also the free-space electric and magnetic energy densities are equal, and, for E_1/E_0 the approximate expression given in Equation E.17b from the energy balance equation (Equation E.18) is obtained.

E.4   ILLUSTRATIVE EXAMPLE

Figure E.1 illustrates the results for an exponential decay profile whose time constant is $(1/b)$:

$$\omega_p^2(t) = \omega_{p0}^2 e^{-bt}. \qquad (E.19)$$

All variables are normalized with reference to the source wave quantities. The source wave frequency ω_0 is taken to be 1 so that ω_1 gives the frequency-upshift ratio. The parameters ω_b and ω_{p0} are assigned the values 100 and 150, respectively. The parameters are such that the Eckersley approximation is valid. A value of 0.01 is assigned to b. The parameter describes a slow decay since $100/2\pi$ cycles of the source wave are accommodated in one time constant. The independent variable, time, is normalized with respect to the period of the source wave.

The results shown in the Figure E.1 for slow decay are numerically computed from Equations E.12, E.14, and E.15, and the following values are obtained at $t = 1000$: $\omega_1/\omega_0 = 15.10$, $E_1/E_0 = 9.96$, $H_1/H_0 = 0.657$, and $S_1/S_0 = 6.57$. Approximate values computed from Equation E.17 give for $t \to \infty$: $\omega_1/\omega_0 \approx n_w \approx 15$, $E_1/E_0 \approx 9.81 \approx 10.6$, $H_1/H_0 \approx 0.654 \approx 0.707$, and $S_1/S_0 \approx 6.38 \approx 7.5$. Here, the first set of approximate values are from the middle set of equations in Equation E.17 and the second set of approximate values are from the last set of equations in E.17.

Figure E.1 also shows, for comparison, horizontal lines that are the results for sudden collapse. These are obtained from Equation E.2b and E.7 and are $n_0 = \omega_1/\omega_0 = 15.11$, $E_1/E_0 = 8.06$, $H_1/H_0 = 0.533$, and $S_1/S_0 = 4.294$. Approximate values obtained from Equation E.9 are $n_w \approx 15$, $E_1/E_0 \approx 7.5$, $H_1/H_0 \approx \frac{1}{2}$, and $S_1/S_0 \approx 3.75$. The approximate results are in agreement with numerically computed results, confirming that Equations E.9 and E.17 can be used as simple approximate expressions under the Eckersley approximation.

E.5   CONCLUSION

The following conclusions can be drawn.

1. The frequency of a source wave propagating in a whistler mode gets upshifted when the ionization collapses. The final upshifted frequency is the source frequency multiplied by the refractive index of the magneto-plasma medium before the collapse begins and is essentially independent of the rate of collapse of the ionization.
2. When the source frequency is much less than the electron gyrofrequency, the refractive index n_w is quite large and the electric field is intensified by a factor of $n_w/2$ for the case of sudden collapse and by a factor of $n_w/\sqrt{2}$ for the case of slow decay. The corresponding intensification factors for the power density are $n_w/4$ and $n_w/2$.

The principle of frequency upshifting using plasmas permits the generation of signals in easily obtainable bands and upshifts them into frequency bands not easily obtainable by other methods. In addition, the output radiation can be tuned by controlling the plasma density. However, the strength of the output radiation is reduced in the process. The technique outlined in this appendix, which involves using a decaying plasma in the presence of a static magnetic field, gives an upshifted as well as intensified signal. Also, the strength of the static magnetic field gives one more controlling parameter for tuning the frequency of the output radiation.

REFERENCES

1. Kalluri, D. K., Effect of switching a magnetoplasma medium on a traveling wave: longitudinal propagation, *IEEE Trans. Antennas Propag.*, AP-37, 1638, 1989.
2. Kalluri, D. K., Frequency shifting using magnetoplasma medium: flash ionization, *IEEE Trans. Plasma Sci.*, 21, 77, 1993.
3. Kalluri, D. K., Goteti, V. R., and Sessler, A. M., WKB solution for wave propagation in a time-varying magnetoplasma medium: longitudinal propagation, *IEEE Trans. Plasma Sci.*, 21, 70, 1993.
4. Goteti, V. R. and Kalluri, D. K., Wave propagation in a switched magnetoplasma medium: transverse propagation, *Radio Sci.*, 25, 61, 1990.
5. Madala, S. R. V. and Kalluri, D. K., Longitudinal propagation of low-frequency waves in a switched magnetoplasma medium, *Radio Sci.*, 28, 121, 1993.
6. Lai, C. H. and Katsouleas, T. C., Frequency upshifting by an ionization front in a magnetized plasma, *IEEE Trans. Plasma Sci.*, 21, 45, 1993.
7. Auld, B. A., Collins, J. H., and Zapp, H. R., Signal processing in a nonperiodically time-varying magnetoelastic medium, *Proc. IEEE*, 56, 258, 1968.
8. Kalluri, D. K., in *Conf. Rec. Abstracts, IEEE Int. Conf. Plasma Science,* Oakland, CA, 1990, 129.
9. Booker, H. G., *Cold Plasma Waves*, Kluwer, Hingham, MA, 1984, 77.
10. Heald, M. A. and Wharton, C. B., *Plasma Diagnostics with Microwaves*, Wiley, New York, 1965, 12.
11. Kalluri, D. K. and Goteti, V. R., Frequency shifting of electromagnetic radiation by sudden creation of a plasma slab, *J. Appl. Phys.*, 72, 4575, 1992.
12. Savage, R. L., Jr., Brogle, R. P., Mori, W. B., and Joshi, C. J., Frequency upshifting and pulse compression via underdense relativistic ionization fronts, *IEEE Trans. Plasma Sci.*, 21, 5, 1993.
13. Lampe, M. and Ott, E., Interaction of electromagnetic waves with a moving ionization front, *Phys. Fluids*, 21, 42, 1978.
14. Lee, J. H. and Kalluri, D. K., Modification of an electromagnetic wave by a time-varying switched magnetoplasma medium: transverse propagation, *IEEE Trans. Plasma Sci.*, 26, 1, 1998.
15. Kalluri, D. K. and Goteti, V. R., WKB solution for wave propagation in a decaying plasma medium, *J. Appl. Phys.*, 66, 3472, 1989.

Appendix F
Conversion of a Whistler Wave into a Controllable Helical Wiggler Magnetic Field*

Dikshitulu K. Kalluri

Abstract — Plasma in the presence of a static magnetic field supports a whistler wave. It is shown that when the static magnetic field is switched off, the energy of the whistler wave is converted into the energy of a helical wiggler magnetic field.

F.1   INTRODUCTION

Coherent radiation generated in a free-electron laser (FEL)[1] is due to the interaction of an electron beam with a wiggler magnetic field, which is a spatially varying static (zero frequency) magnetic field. Linear or helical wigglers can be constructed from electromagnets or permanent magnets. The B field of a helical wiggler can be written as

$$\overline{B}_W(z) = \left(j\hat{x} + \hat{y} \right) B_W \cos\left(k_W z \right), \tag{F.1}$$

where k_W is the wave number of the wiggler. The wavelength λ of the output radiation is given by:[1]

$$\lambda = \frac{\lambda_W}{2\gamma_0^2} \left(1 + \frac{K_W^2}{2} \right), \tag{F.2a}$$

where

$$\lambda_W = \frac{2\pi}{k_W}, \tag{F.2b}$$

* © American Institute of Physics. Reprinted with permission from Kalluri, D. K., *Journal of Applied Physics*, 79, 6770–6774, 1996.

$$K_W = \frac{qB_W}{\left(mck_W\right)}, \tag{F.2c}$$

$$\gamma_0 = \left(1 - \frac{v_0^2}{c^2}\right)^{1/2}. \tag{F.2d}$$

Here, λ_W is the wiggler wavelength (spatial period), K_W is the wiggler strength parameter, q is the absolute value of the charge of the electron, v_0 is the velocity of the electron beam, and the other symbols have the usual meaning.[1] The tunability of the FEL comes from the variability of the parameter γ_0 through the kinetic energy of the electrons. Once a FEL is constructed, λ_W is fixed.

It is known[2-5] that part of the energy of a source wave is converted into a wiggler magnetic field when an unbounded plasma medium is suddenly switched on. If the source wave in free space is of right circular polarization given by

$$\bar{e}_i(z,t) = (\hat{x} - j\hat{y})E_0 \exp\left[j\left(\omega_0 t - k_0 z\right)\right], \tag{F.3a}$$

$$\bar{h}_i(z,t) = (j\hat{x} + \hat{y})H_0 \exp\left[j\left(\omega_0 t - k_0 z\right)\right], \tag{F.3b}$$

$$k_0 = \omega_0/c = 2\pi/\lambda_0, \tag{F.3c}$$

the wiggler field produced is a helical magnetic field given by

$$\frac{H_W}{H_0} = \frac{\omega_p^2}{\omega_0^2 + \omega_p^2} \cos\left(k_0 z\right), \tag{F.4a}$$

$$B_W = \mu_0 H_W. \tag{F.4b}$$

Here, ω_0 is the angular source frequency, ω_p is the plasma frequency of the switched medium, and λ_0 is the wavelength in free space of the source wave, which is also the period of the generated wiggler field. For a fixed source wave, the wiggler period is also fixed.

This appendix deals with the principle of establishing a helical wiggler magnetic field with controllable λ_W even if the source wave is of fixed frequency. A whistler wave that propagates in a magnetoplasma when the source frequency ω_0 is less than the electron gyrofrequency ω_b gives rise to a wiggler magnetic field in the plasma when the static magnetic field is switched off. The strength and the period of the wiggler field depend on the parameters of the magnetoplasma medium. When the parameters are such that ω_0 is much less than ω_b and ω_p, the total energy of the source wave is converted into the magnetic energy of the wiggler field.

F.2   FORMULATION

A wave with right circular polarization, called an R wave,[6,7] is assumed to be propagating in the positive z-direction in a spatially unbounded magnetoplasma medium. The electric, magnetic, and velocity fields of this wave, named the source wave, are

$$\bar{e}_i(z,t) = (\hat{x} - j\hat{y})E_0 \exp\left[j(\omega_0 t - k_W z)\right],$$ (F.5a)

$$\bar{h}_i(z,t) = (j\hat{x} + \hat{y})H_0 \exp\left[j(\omega_0 t - k_W z)\right],$$ (F.5b)

$$\bar{v}_i(z,t) = (\hat{x} - j\hat{y})V_0 \exp\left[j(\omega_0 t - k_W z)\right],$$ (F.5c)

where ω_0 is the frequency of the incident wave and

$$k_W = \omega_0(n_R/c),$$ (F.5d)

$$n_R = \sqrt{1 + \frac{\omega_p^2}{\omega_0(\omega_{b0} - \omega_0)}},$$ (F.5e)

$$\eta_0 H_0 = n_R E_0,$$ (F.5f)

$$(m/q)V_0 = \frac{j}{\omega_0 - \omega_b} E_0.$$ (F.5g)

Here, c is the velocity of light in free space, η_0 is the intrinsic impedance of free space, ω_p is the plasma frequency, $\omega_{b0} = qB_{s0}/m$ is the electron gyrofrequency, and n_R is the refractive index of the magnetoplasma medium existing prior to $t = 0$.

F.3   SUDDEN SWITCHING-OFF OF THE STATIC MAGNETIC FIELD

Assume that at $t = 0$ the static magnetic field B_{s0} is switched off in a short time. Let us idealize the problem by considering a sudden switching off of the static magnetic field. The fields for $t > 0$ can be expressed as

$$\bar{e}(z,t) = (\hat{x} - j\hat{y})e(t)\exp\left[-jk_W z\right], \qquad t > 0,$$ (F.6a)

$$\bar{h}(z,t) = (j\hat{x} + \hat{y})h(t)\exp\left[-jk_W z\right], \qquad t > 0,$$ (F.6b)

$$\bar{\upsilon}(z,t) = (\hat{x} - j\hat{y})\upsilon(t)\exp[-jk_W z], \qquad t > 0. \tag{F.6c}$$

In writing the above expressions, the well-known condition[2,8,9] that the wave number is conserved at a time discontinuity in the properties of the medium is used. From Maxwell's equations and the force equation for the velocity field, one obtains:

$$\frac{d}{dt}(\eta_0 h) = (j\omega_0 n_R)e, \quad t > 0, \tag{F.7a}$$

$$\frac{d}{dt}(e) = (j\omega_0 n_R)(\eta_0 h) + \omega_p^2 u, \quad t > 0, \tag{F.7b}$$

$$\frac{d}{dt}(u) = -e, \quad t > 0, \tag{F.7c}$$

where $u = m\upsilon/q$.

F.3.1   FREQUENCIES AND THE FIELDS OF THE NEW WAVES

Equations F.7, subject to the initial values $e(0) = E_0$, $h(0) = H_0$, and $u(0) = mV_0/q$, are solved using the Laplace transform technique. The following solutions are obtained for the frequencies and fields of the new waves:

$$e(t) = \sum_1^3 E_n \exp(j\omega_n t), \tag{F.8a}$$

$$h(t) = \sum_1^3 H_n \exp(j\omega_n t), \tag{F.8b}$$

$$\omega_1 = \omega_{up}, \quad \omega_2 = -\omega_{up}, \quad \omega_3 = 0, \tag{F.8c}$$

$$\frac{E_1}{E_0} = (1 + \omega_0/\omega_{up})/2, \quad \frac{E_2}{E_0} = (1 - \omega_0/\omega_{up})/2, \quad \frac{E_3}{E_0} = 0, \tag{F.8d}$$

$$\frac{H_1}{H_0} = \frac{\omega_0}{\omega_{up}}\frac{E_1}{E_0}, \quad \frac{H_2}{H_0} = -\frac{\omega_0}{\omega_{up}}\frac{E_2}{E_0}, \quad \frac{H_3}{H_0} = \left(1 - \frac{\omega_0^2}{\omega_{up}^2}\right), \tag{F.8e}$$

$$\frac{S_1}{S_0} = \frac{\omega_0}{4\omega_{up}}(1 + \omega_0/\omega_{up})^2, \quad \frac{S_2}{S_0} = -\frac{\omega_0}{4\omega_{up}}(1 - \omega_0/\omega_{up})^2, \quad \frac{S_3}{S_0} = 0, \tag{F.8f}$$

where

$$\omega_{up} = \sqrt{\omega_p^2 + n_R^2 \omega_0^2}.$$ (F.8g)

Here, a negative value for ω indicates a reflected wave traveling in a direction opposite to that of the source wave and a zero value for ω gives a wiggler field. The symbol S stands for power density equal to EH.

F.3.2 DAMPING RATES

The above calculations are made assuming the plasma is lossless. The modification for a lossy plasma is derived next by a simple modification of Equation F.7c,

$$\frac{d}{dt}(u) = -e - \nu u, \qquad t > 0.$$ (F.9)

Here ν is the collision frequency. For a low-loss case, ν is small and an approximate solution is obtained using the Laplace transform technique.[10] The results are approximately the same as those given by Equation F.8 except the fields are damped because of the losses in the plasma. The damping time constants for the waves are given by

$$t_n = \frac{\left(3\omega_n^2 - \omega_p^2 - n_R^2 \omega_0^2\right)}{\nu\left(\omega_n^2 - n_R^2 \omega_0^2\right)}, \qquad n = 1, 2, 3.$$ (F.10)

The wiggler field is produced by the third wave for which $\omega_n = \omega_3 = 0$.

F.3.3 ECKERSLEY APPROXIMATION

The refractive index $n_R = n_w$ is large when $\omega_0 \ll \omega_{b0} \sim \omega_p$ and from Equation F.5e

$$n_R \approx n_w = \omega_{p0}/\sqrt{\omega_0 \omega_{b0}}, \qquad \omega_0 \ll \omega_{b0} \sim \omega_p,$$ (F.11a)

and from Equation F.8,

$$\omega_1 = \omega_{up} \approx \omega_p, \qquad \omega_2 = -\omega_{up} \approx -\omega_p, \qquad \omega_3 = 0,$$ (F.11b)

$$E_1/E_0 \approx 1/2, \qquad E_2/E_0 \approx 1/2, \qquad E_3/E_0 = 0,$$ (F.11c)

$$H_1/H_0 \approx 0, \qquad H_2/H_0 \approx 0, \qquad H_3/H_0 \approx 1,$$ (F.11d)

$$S_1/S_0 \approx 0, \qquad S_2/S_0 \approx 0, \qquad S_3/S_0 = 0.$$ (F.11e)

From Equation F.10, the Eckersley approximations for the damping constants can be obtained,

$$t_{1,2} \approx \frac{2}{v\left(1+\omega_0/\omega_{b0}\right)} \approx \frac{2}{v}, \tag{F.11f}$$

$$t_3 \approx \frac{1}{v}\frac{\omega_{b0}}{\omega_0}\left(1+\frac{\omega_0}{\omega_{b0}}\right) \approx \frac{1}{v}\frac{\omega_{b0}}{\omega_0}. \tag{F.11g}$$

The wiggler magnetic field decays in the plasma much more slowly than the wave fields. From Equation F.11 it is clear that the first and second waves with upshifted frequencies are weak, but the third mode, which is a wiggler magnetic field, is quite strong. The energy in the whistler wave, under Eckersley approximation, is predominantly the magnetic energy due to the plasma current, and its density w_m is given by[6]

$$w_m = \tfrac{1}{2}\varepsilon_0 E_0^2 n_w^2. \tag{F.12}$$

After the static magnetic field B_s collapses, the upshifted first and second waves are weak and do not carry much energy. Most of the energy for $t > 0$ is in the wiggler magnetic field H_3. Thus,

$$\tfrac{1}{2}\varepsilon_0 E_0^2 n_w^2 \approx \tfrac{1}{2}\mu_0 H_3^2, \tag{F.13a}$$

$$H_3 \approx n_w \frac{E_0}{\eta_0} = H_0. \tag{F.13b}$$

The wiggler strength parameter K_W and λ_W can now be approximated as

$$K_W \approx \frac{qE_0}{(mc\omega_0)} = \frac{586.23E_0}{\omega_0}, \quad \omega_0 \ll \omega_{b0} \sim \omega_p, \tag{F.14a}$$

$$\lambda_W \approx \frac{\lambda_0}{n_w}, \quad \omega_0 \ll \omega_{b0} \sim \omega_p. \tag{F.14b}$$

The following sample calculation illustrates the tunability that can be achieved. Let the source wave be of frequency $f_0 = 1$ GHz so that $\lambda_0 = 30$ cm and $f_p = 100$ GHz. Let the dc current in the magnet be varied so that f_{b0} can be fixed from 10 to 50 GHz. The refractive index n_w changes from 100/3 to 100/7, and the wiggler wavelength λ_W changes from 9 to 2.1 cm.

F.4 SLOW DECAY OF B_S

The formulation of Section F.3 assumed the sudden collapse of B_s, which is unrealistic. If the dc current of the magnet is switched off, due to inductance, B_s will not decline suddenly, but will decay over a period of time. In this section the effect of a slow decline of B_s will be studied. A Wentzel–Kramers–Brillouin (WKB) type of solution can be obtained for the case of slow decay; i.e., $\omega_b(t) = qB_s(t)/m$ is a slowly decaying function of time. The differential equations satisfied by the fields h and e are given by Equations F.7a and b, respectively, and the differential equation for u is given by

$$\frac{d}{dt}(u) = -e + j\omega_b(t)u, \quad t > 0. \tag{F.15}$$

From Equations F.7a and b and Equation F.15, a third-order differential equation for h can be obtained,

$$\frac{d^3h}{dt^3} - j\omega_b(t)\frac{d^2h}{dt^2} + \left[\omega_0^2 n_R^2 + \omega_p^2\right]\frac{dh}{dt} - j\omega_0^2 n_R^2 \omega_b(t)h = 0. \tag{F.16}$$

The technique of solving this equation is similar to that used in References 11 through 13, and, therefore, much of the mathematical detail is omitted here. A complex instantaneous frequency function is defined such that

$$\frac{dh(t)}{dt} = p(t)h(t) = \left[\alpha(t) + j\omega(t)\right]h(t). \tag{F.17}$$

Here, $\omega(t)$ is the instantaneous frequency. By substituting Equation F.17 in Equation F.16 and neglecting α and all derivatives, a zeroth-order solution can be obtained. The solution is a cubic in ω, giving the instantaneous frequencies of three waves created by the switching action,

$$\omega^3 - \omega_b(t)\omega^2 - \left(n_R^2\omega_0^2 + \omega_p^2\right)\omega + n_R^2\omega_0^2\omega_b(t) = 0. \tag{F.18}$$

This cubic has two positive real roots and one negative real root and has the following solutions at $t = 0$ and $t = \infty$:

$$\omega_1(0) = \sqrt{\left(\frac{\omega_0 + \omega_b(0)}{2}\right)^2 + \frac{\omega_p^2\omega_b(0)}{\omega_b(0) - \omega_0}} + \frac{-\omega_0 + \omega_b(0)}{2}, \tag{F.19a}$$

$$\omega_2(0) = -\sqrt{\left(\frac{\omega_0 + \omega_b(0)}{2}\right)^2 + \frac{\omega_p^2\omega_b(0)}{\omega_b(0) - \omega_0}} + \frac{-\omega_0 + \omega_b(0)}{2}, \tag{F.19b}$$

$$\omega_3(0) = \omega_0, \tag{F.19c}$$

$$\omega_1(\infty) = \omega_{up}, \tag{F.19d}$$

$$\omega_2(\infty) = -\omega_{up}, \tag{F.19e}$$

$$\omega_3(\infty) = 0. \tag{F.19f}$$

An equation for α can now be obtained by substituting Equation F.17 in Equation F.16 and equating the real part to zero. In obtaining the WKB solution, the derivatives and powers of α, etc. are neglected:

$$\alpha = \dot{\omega}\,\frac{3\omega - \omega_b}{\omega_0^2 n_R^2 + \omega_p^2 - 3\omega^2 + 2\omega\omega_b}. \tag{F.20}$$

The amplitudes of the fields associated with these waves are obtained by solving the initial value problem. The continuity of the electric and magnetic fields provides two initial conditions. The third initial condition is the continuity of the velocity field at $t = 0$. The amplitudes of the magnetic fields of two of the waves are of the order of the time derivative of $\omega_b(t)$ curve and negligible. The amplitude of the magnetic field of the wave whose initial frequency is ω_0 is significant, and this wave is called a modified source wave. Its magnetic field $h_3(z,t)$ can be expressed as

$$h_3(z,t) = H_3(t)\exp\left[j\left(\int_0^t \omega_3(t)dt - k_W z\right)\right], \tag{F.21}$$

where

$$H_3(t) = H_0 \exp\left(\int_0^t \alpha_3(\tau)d\tau\right). \tag{F.22}$$

The integral in Equation F.22 can be evaluated numerically, but in this case further mathematical manipulation will permit an analytical integration. From Equation F.18

$$\omega_b = \omega\left[1 - \frac{\omega_p^2}{\omega^2 - n_0^2\omega_0^2}\right]. \tag{F.23}$$

By substituting for ω_b in Equation F.20,

$$\alpha = \dot{\omega}\,\frac{-\omega\left(2\omega^2 - 2n_R^2\omega_0^2 + \omega_p^2\right)}{\left(\omega^2 - n_R^2\omega_0^2\right)^2 + \omega_p^2\omega^2 + n_R^2\omega_0^2\omega_p^2}. \tag{F.24}$$

By denoting the denominator in Equation F.24 by a variable r,

$$\alpha = -\frac{1}{2r}\frac{dr}{d\omega}\dot{\omega},$$
(F.25)

$$\int_0^t \alpha(\tau)d\tau = -\frac{1}{2}\int_{r(0)}^{r(t)}\frac{dr}{r} = \ln\sqrt{\frac{r[\omega(0)]}{r[\omega(t)]}}.$$
(F.26)

Therefore,

$$H_3(t) = H_0\sqrt{\frac{\left(\omega_3^2(0)-n_R^2\omega_0^2\right)^2+\omega_p^2\omega_3^2(0)+n_R^2\omega_0^2\omega_p^2}{\left(\omega_3^2(t)-n_R^2\omega_0^2\right)^2+\omega_p^2\omega_3^2(t)+n_R^2\omega_0^2\omega_p^2}}.$$
(F.27)

For the third wave $\omega_3(0) = \omega_0$ and Equation F.27 becomes

$$H_3(t) = H_0\sqrt{\frac{\left(n_R^2-1\right)^2+\left(\omega_p^2/\omega_0^2\right)\left(n_R^2+1\right)}{\left(n_R^2-\left[\omega_3^2(t)/\omega_0^2\right]\right)^2+\left(\omega_p^2/\omega_0^2\right)\left\{n_R^2+\left[\omega_3^2(t)/\omega_0^2\right]\right\}}},$$
(F.28a)

$$E_3(t) = E_0\frac{\omega_3(t)}{\omega_0}.$$
(F.28b)

Since $\omega_3(\infty) = 0$,

$$H_3(\infty) = H_0\sqrt{\frac{\left(n_R^2-1\right)^2+\left(\omega_p^2/\omega_0^2\right)\left(n_R^2+1\right)}{n_R^4+\left(\omega_p^2/\omega_0^2\right)n_R^2}},$$
(F.29a)

$$E_3(\infty) = 0.$$
(F.29b)

For Eckersley approximation $n_R \gg 1$, and

$$H_3(\infty) = H_0, \quad \omega_0 \ll \omega_{b0} \sim \omega_p,$$
(F.30a)

$$E_3(\infty) = 0.$$
(F.30b)

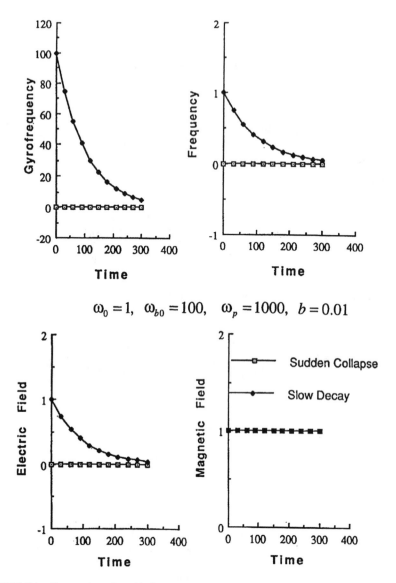

$$\omega_0 = 1, \quad \omega_{b0} = 100, \quad \omega_p = 1000, \quad b = 0.01$$

FIGURE F.1 Conversion of a whistler wave into a helical magnetic wiggler field. The values on the vertical axes are normalized with respect to source wave quantities. The source frequency ω_0 is taken as 1. The horizontal timescale is normalized with respect to the period of the source wave. The parameters are such that the Eckersley approximation is applicable.

F.5   ILLUSTRATIVE EXAMPLE

Figure F.1 illustrates the results for an exponential decay profile whose time constant is $(1/b)$,

$$\omega_b(t) = \omega_{b0}e^{-bt}. \tag{F.31}$$

All variables are normalized with reference to the source wave quantities. The source wave frequency ω_0 is taken to be 1. The parameters ω_{b0} and ω_p are assigned the values 100 and 1000, respectively. The parameters are such that Eckersley approximation is valid. A value of 0.01 is assigned to b. The parameter describes a slow decay since $100/2\pi$ cycles of the source wave are accommodated in one time constant. The independent variable, time, is normalized with respect to the period of the source wave.

The results shown in Figure F.1 for slow decay are numerically computed from Equations F.18 and F.28a and b, and the following values are obtained at $t = 1000$: $\omega_b/\omega_0 = 4.54 \times 10^{-3}$, $\omega_3/\omega_0 = 4.54 \times 10^{-5}$, $E_3/E_0 = 4.54 \times 10^{-5}$, and $H_3/H_0 = 1.000048$. Figure F.1 also shows, for comparison, horizontal lines that are the results for sudden collapse. Irrespective of the rate of collapse of B_s, a strong wiggler field is generated.

Figure F.2 shows results for $\omega_{b0} = 1.1$ and $\omega_p = 0.5$ with the other parameters remaining the same as those of Figure F.1. The Eckersley approximation is not valid in this case, but Equation F.28 still holds good; the third wave becomes the wiggler magnetic field with $H_3/H_0 = 0.75$; however, the damping time constant given by Equation F.10 is much less than the value for the parameters of Figure F.1.

F.6   CONCLUSION

The effect of switching off of the static magnetic field on a whistler wave is considered. A sudden collapse as well as a slow decay of the static magnetic field showed that the energy of the whistler wave is converted into a wiggler magnetic field.

ACKNOWLEDGMENT

The author thanks the ionospheric effects division of the Phillips Laboratory at Hanscom for providing him an opportunity to complete this work during the summer of 1995.

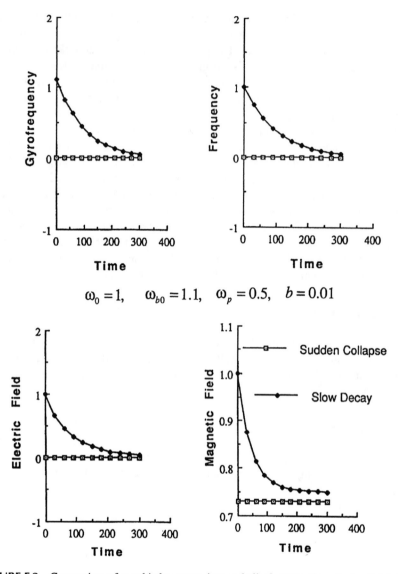

$$\omega_0 = 1, \quad \omega_{b0} = 1.1, \quad \omega_p = 0.5, \quad b = 0.01$$

FIGURE F.2 Conversion of a whistler wave into a helical magnetic wiggler field. The parameters are such that Eckersley approximation is not applicable.

REFERENCES

1. Granastein, V. L. and Alexeff, I., *High-Power Microwave Sources*, Artech House, Boston, 1987, 207.
2. Jiang, C. L., Wave propagation and dipole radiation in a suddenly created plasma, *IEEE Trans. Antennas Propag.*, AP-23, 83, 1975.
3. Wilks, S. C., Dawson, J. M., and Mori W. B., Frequency up-conversion of electromagnetic radiation with use of an overdense plasma, *Phys. Rev. Lett.*, 63, 337, 1988.
4. Kalluri, D. K., Effect of switching a magnetoplasma medium on a traveling wave: longitudinal propagation, *IEEE Trans. Antennas Propag.*, AP-37, 1638, 1989.
5. Lapierre, D. A., Master's Thesis, University of Massachusetts Lowell, 1993.
6. Booker, H. G., *Cold Plasma Waves*, Kluwer, Hingham, MA, 1984, 77.
7. Heald, M. A. and Wharton, C. B., *Plasma Diagnostics with Microwaves*, Wiley, New York, 1965, 12.
8. Auld, B. A., Collins, J. H., and Zapp, H. R., Signal processing in a nonperiodically time-varying magnetoelastic medium, *Proc. IEEE*, 56, 258, 1968.
9. Kalluri, D. K., in *Conf. Rec. Abstracts, IEEE Int. Conf. Plasma Science,* Oakland, CA, 1990, 129.
10. Kalluri, D. K. and Goteti, V. R., Damping rates of waves in a switched magnetoplasma medium: longitudinal propagation, *IEEE Trans. Plasma Sci.*, 18, 797, 1990.
11. Kalluri, D. K., Goteti, V. R., and Sessler, A. M., WKB solution for wave propagation in a time-varying magnetoplasma medium: longitudinal propagation, *IEEE Trans. Plasma Sci.*, 21, 70, 1993.
12. Lee, J. H. and Kalluri, D. K., Modification of an electromagnetic wave by a time-varying switched mangetoplasma medium: transverse propagation, *IEEE Trans. Plasma Sci.*, 26, 1998.
13. Kalluri, D. K. and Goteti, V. R., WKB solution for wave propagation in a decaying plasma medium, *J. Appl. Phys.*, 66, 3472, 1989.

Appendix G
Effect of Switching a
Magnetoplasma Medium
on the Duration of a
Monochromatic Pulse*

Dikshitulu K. Kalluri

Abstract — The main effect of switching a magnetoplasma medium is to split the source wave into new waves whose frequencies are different from the source wave. In addition, if the source is a monochromatic pulse, the duration of the pulse is altered. Analytical expressions for the pulse duration of the various characteristic waves in a magnetoplasma are derived. The variations of the pulse duration with the source frequency, the plasma frequency, and the cyclotron frequency are illustrated. The principle of the change in the pulse duration can be used to diagnose dynamically the time-varying parameters of a magnetoplasma medium.

G.1 INTRODUCTION

The study of the interaction of electromagnetic waves with time-varying plasmas has been a subject of considerable recent activity in plasma science[1] because of its potential applications in the generation of tunable electromagnetic radiation over a broad frequency range.

Jiang[2] investigated the effect of switching an unbounded isotropic cold plasma medium suddenly on a monochromatic plane electromagnetic wave. The wave propagation in such a medium is governed by the conservation of the wave number across the time discontinuity and the consequent change in the wave frequency. Rapid creation of the medium can be approximated as a sudden switching of the medium. The main effect of switching the medium is the splitting of the original wave (henceforth called incident wave, in the sense of incidence on a time discontinuity in the properties of the medium) into new waves whose frequencies are different from the incident wave. Recent experimental work[3-5] and computer simulation[6]

* © 1997 Plenum Publishing Corporation. Reprint with permission from Kalluri, D. K., *International Journal of Infrared and Millimeter Waves*, 18, 1585–1603, 1997.

demonstrated the frequency upshifting of microwave radiation by rapid plasma creation.

G.2 EFFECT ON A MONOCHROMATIC WAVE: REVIEW

It is known[7] that the presence of a static magnetic field will influence the frequency shift, etc. Rapid creation of a plasma medium in the presence of a static magnetic field can be approximated as the sudden switching of a magnetoplasma medium. An idealized mathematical description of the problem is given next. A uniform plane wave is traveling in free space along the z-direction, and it is assumed that, at $t = 0^-$, the wave occupies the entire space. A static magnetic field of strength B_0 is assumed to be present throughout. At $t = 0$, the entire space is suddenly ionized, thus converting the medium from free space to a magnetoplasma medium. The incident wave frequency is assumed to be high enough that the effects of ion motion may be neglected.

The strength and the direction of the static magnetic field affect the number of new waves created, their frequencies, and the power density of these waves. Some of the new waves are transmitted waves (waves propagating in the same direction as the incident wave), and some are reflected waves (waves propagating in the opposite direction to that of the incident wave). Reflected waves tend to have less power density.

A physical interpretation of the waves can be given in the following way. The electric and magnetic fields of the incident wave and the static magnetic field accelerate the electrons in the newly created magnetoplasma; the electrons in turn radiate new waves. The frequencies of the new waves and their fields can be obtained by adding contributions from the many electrons whose positions and motions are correlated by the collective effects supported by the magnetoplasma medium. Such a detailed calculation of the radiated fields seems to be quite involved. A simple, but less accurate, description of the plasma effect is obtained by modeling the magnetoplasma as a dielectric medium whose refractive index n, for the type of wave propagation, is computed through magnetoionic theory.[8] The frequency of the new waves is constrained by the requirements that the wave number (k_0) is conserved over the time discontinuity. This gives a conservation[9] law, $k_0 c = \omega_0 = n(\omega)\omega$, from which ω can be determined. Solution of the associated electromagnetic initial value problem gives the electric and magnetic fields of the new waves. By using this approach, the general aspects of wave propagation in a switched magnetoplasma medium were discussed earlier[7,10,11] for the cases of L, R, and X incidence.[8]

L incidence is an abbreviation used to describe the case where the incident wave has left-hand circular polarization and the static magnetic field is in the z-direction (Figure G.1a). Three new waves with left-hand circular polarization labeled $L1$, $L2$, and $L3$ are generated by the medium switching. $L1$ is a transmitted wave. R incidence refers to the case where the incident wave has right-hand circular polarization and the static magnetic field is in the z-direction. The medium switching, in this case, creates three R waves labeled as $R1$, $R2$, and $R3$. $R1$ and $R3$ are transmitted waves

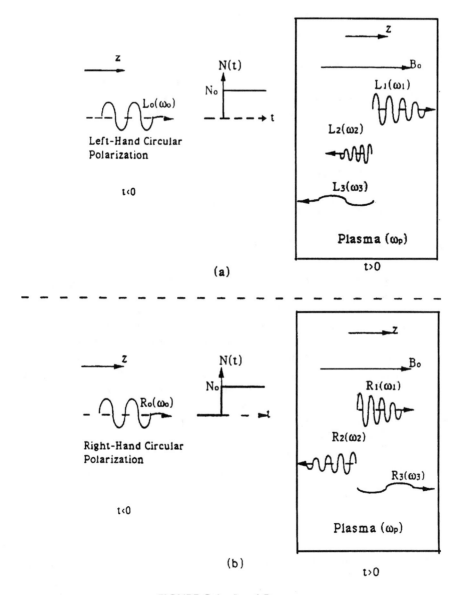

FIGURE G.1 *L* and *R* waves.

(Figure G.1b). The frequencies of the *R* waves are the roots of the cubic characteristic equation:[7]

$$\omega_n^3 - \omega_b\omega_n^2 - \left(\omega_0^2 + \omega_p^2\right)\omega_n + \omega_0^2\omega_b = 0. \qquad (G.1)$$

In the above, ω_b is the electron gyrofrequency and the incident wave variables are denoted by the zero subscript. A negative value for ω_n^7 indicates that the wave is a reflected wave, i.e., that it propagates in a direction opposite to that of the source (incident) wave. If the source wave is right going, then the reflected wave will be left going.

X incidence refers to the case where the electric field of the incident wave is in the x-direction and the static magnetic field is in the y-direction. The medium switching in this case creates four extraordinary waves labeled $X1$, $X2$, $X3$, and $X4$. $X1$ and $X2$ are transmitted waves. The frequencies of these waves are the roots of the characteristic equation:[10]

$$\omega_n^4 - \omega_n^2\left(\omega_0^2 + \omega_b^2 + 2\omega_p^2\right) + \left(\omega_p^4 + \omega_0^2\left(\omega_b^2 + \omega_p^2\right)\right) = 0. \tag{G.2}$$

The source in an experiment is likely to be a monochromatic pulse of a given duration. In this appendix, the effect of switching a magnetoplasma medium on the duration of such a pulse is considered.

G.3 EFFECT ON THE DURATION OF A MONOCHROMATIC PULSE

Let ω_0 be the frequency, k the wave number, L_0 the length, T_0 the duration of the monochromatic pulse traveling in the z-direction in free space (Figure G.2a). At $t = 0$, let the medium be converted to a magnetoplasma whose parameters are ω_b and ω_p. The effects of these parameters on the frequencies and amplitudes of the new pulses will be the same as those on the monochromatic plane wave and are discussed in References 7, 10, and 11. At a temporal discontinuity, the pulse length L_0 is conserved. However, the pulse duration is altered. A geometrical interpretation of the change in the pulse duration for O, R, and X pulses is given in Figures G.2b, G.3, and G.4, respectively. The effect on the duration of the L pulses is easily inferred from the results for the R pulses.

The pulse duration and the pulse length for the nth wave are related by

$$T_n = \frac{L_0}{v_{grn}}. \tag{G.3}$$

where v_{grn} is the group velocity of the nth wave. For the source pulse in free space, the group velocity is c and

$$T_0 = L_0/c. \tag{G.4}$$

From the dispersion relation, the group velocity of the nth wave ($d\omega_n/dk$) can be obtained. The pulse duration ratio is given by

$$\frac{T_n}{T_0} = \frac{c}{\left|v_{grn}\right|} = \frac{c}{\left|d\omega_n/dk\right|} = \frac{1}{\left|d\omega_n/d\omega_0\right|}. \tag{G.5}$$

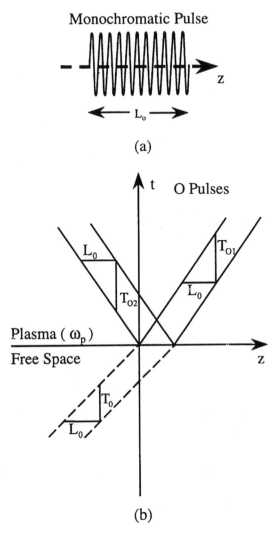

FIGURE G.2   (a) Monochromatic source pulse in free space; (b) duration of O pulses in plasma.

An increase (decrease) in the numerical value of the group velocity produces a decrease (increase) in the pulse duration.

For O waves in an isotropic plasma medium, the dispersion relation is

$$\omega_n^2 = \omega_0^2 + \omega_p^2 \tag{G.6}$$

and

$$\frac{T_{O1,O2}}{T_0} = \frac{\omega_0}{|\omega_n|}. \tag{G.7}$$

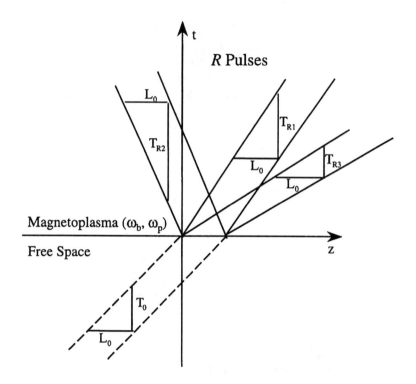

FIGURE G.3   Duration of R pulses.

For R waves, the dispersion relation is given by Equation G.1 and

$$\frac{d\omega_n}{d\omega_0} = \frac{2\omega_0(\omega_n - \omega_0)}{3\omega_n^2 - 2\omega_n\omega_b - \omega_0^2\omega_p^2}. \tag{G.8}$$

For X waves, the dispersion relation is given by Equation G.2 and

$$\frac{d\omega_n}{d\omega_0} = \frac{\omega_0}{\omega_n}\frac{\omega_n^2 - \omega_b^2 - \omega_p^2}{2\omega_n^2 - \left(\omega_0^2 + \omega_b^2 + 2\omega_p^2\right)}. \tag{G.9}$$

G.4 NUMERICAL RESULTS AND DISCUSSION

The top portions of Figures G.5 through G.9 show the variation of frequency of the pulses as a function of the frequency ω_0 of the source pulse in free space. The frequencies are normalized by choosing the plasma frequency parameter $\omega_p = 1.0$. The gyrofrequency parameter ω_b is varied from 0 to 1.5 in steps of 0.5. The slope of the pulse frequency vs. the source frequency curve is proportional to the group velocity, and the pulse duration is inversely proportional to the magnitude of this slope. The bottom portions of these curves describe the variation of the pulse duration

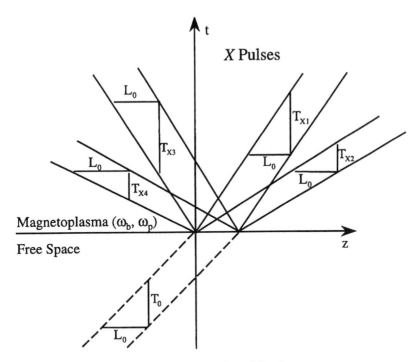

FIGURE G.4 Duration of X pulses.

of the various B pulses. Pulse duration is normalized by choosing $T_0 = 1$. The $R1$ and $X1$ curves for $\omega_b = 0$ describe the $O1$ pulse. The following interesting points can be noted:

1. Pulse duration ratio for $R1$ and $X1$ pulses asymptotically reaches unit value as ω_0 tends to infinity. For large values of ω_0, the plasma has very little effect.
2. $R2$ is a reflected pulse and its graph for the pulse duration is qualitatively similar to the $R1$ graph. $O2$, $X3$, $X4$ pulses are also reflected pulses, and their graphs are the same as those for $O1$, $X1$, $X2$ pulses, respectively.
3. $R3$ and $X2$ pulses are generated only in a switched anisotropic plasma medium, and their behavior is strongly influenced by the ω_b parameter. Their pulse duration goes through a minimum, and the minimum point is strongly influenced by the ω_b parameter.

It is suggested that the principle of the change in the pulse duration can be used to diagnose dynamically the time-varying parameters of an anisotropic medium.

ACKNOWLEDGMENT

The work is supported by Air Force Office of Scientific Research under contract AFOSR 97-0847.

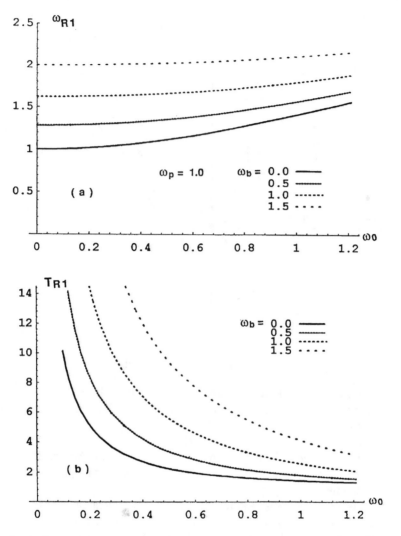

FIGURE G.5 $R1$ pulse: (a) pulse frequency vs. source frequency; (b) pulse duration vs. source frequency.

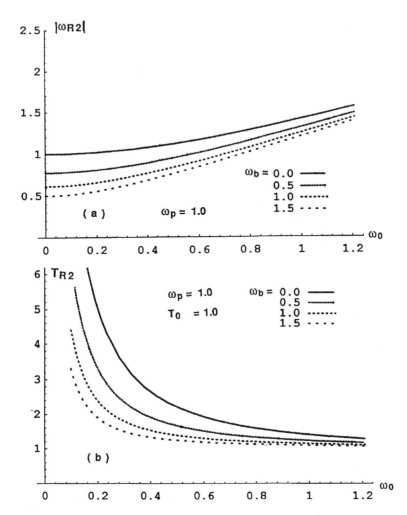

FIGURE G.6 *R*2 pulse: (a) pulse frequency vs. source frequency; (b) pulse duration vs. source frequency.

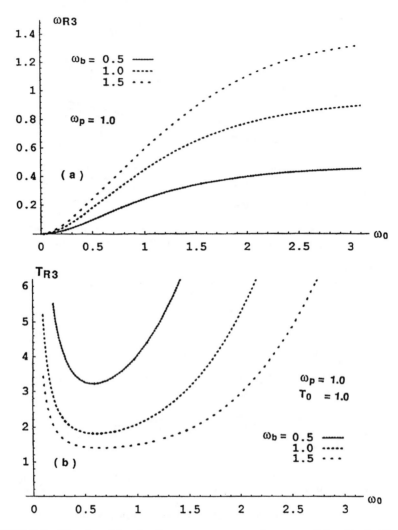

FIGURE G.7 *R*3 pulse: (a) pulse frequency vs. source frequency; (b) pulse duration vs. source frequency.

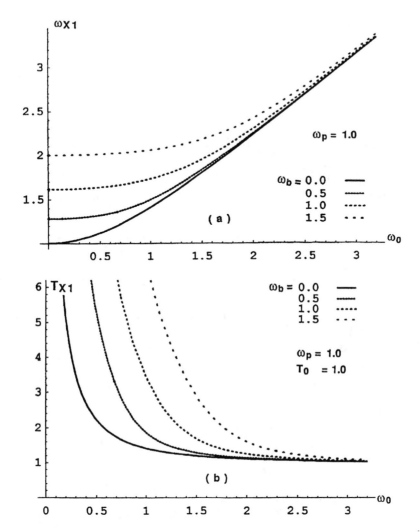

FIGURE G.8 *X*1 pulse: (a) pulse frequency vs. source frequency; (b) pulse duration vs. source frequency.

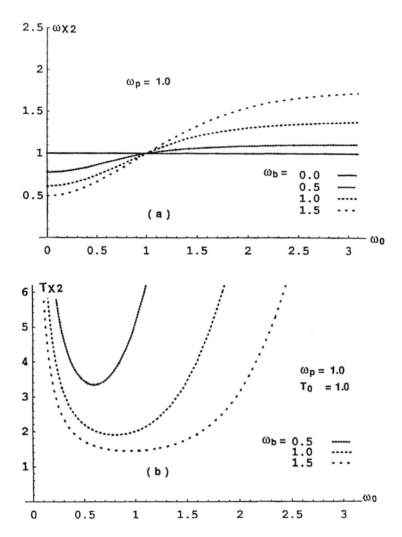

FIGURE G.9 *X*2 pulse: (a) pulse frequency vs. source frequency; (b) pulse duration vs. source frequency.

REFERENCES

1. Mori, W. B., Ed., *IEEE Trans. Plasma Sci.*, 21, 1, 1993.
2. Jiang, C. L., Wave propagation and dipole radiation in a suddenly created plasma, *IEEE Trans. Antennas Propag.*, AP-23, 83, 1975.
3. Rader, M. and Alexeff, I., *Int. J. Infrared Millimeter Waves*, 12, 683, 1991.
4. Kuo, S. P., Zhang, Y. S., and Ren, A. Q., in *Conf. Rec. Abstr., IEEE Int. Conf. Plasma Science,* Oakland, CA, 1990, 171.
5. Joshi, C. J., Clayton, C. E., Marsh, K., Hopkins, D. B., Sessler, A., and Whittum, D., Demonstration of the frequency upshifting of microwave radiation by rapid plasma creation, *IEEE Trans. Plasma Sci.*, 18, 814, 1990.
6. Wilks, S. C., Dawson, J. M., and Mori, W. B., Frequency up-conversion of electromagnetic radiation with use of an overdense plasma, *Phys. Rev. Lett.*, 61, 337, 1988.
7. Kalluri, D. K., Effect of switching a magnetoplasma medium on a traveling wave: longitudinal propagation, *IEEE Trans. Antennas Propag.*, AP-37, 1638, 1989.
8. Booker, H. G., *Cold Plasma Waves*, Kluwer, Hingham, MA, 1984.
9. Kalluri, D. K., in *Conf. Rec. Abstr., IEEE Int. Conf. Plasma Science,* Oakland, CA, 1990, 129.
10. Goteti, V. R. and Kalluri, D. K., Wave propagation in a switched magnetoplasma medium: transverse propagation, *Radio Sci.*, 25, 61, 1990.
11. Kalluri, D. K., Frequency shifting using magnetoplasma medium: flash ionization, *IEEE Trans. Plasma Sci.*, 21, 77, 1993.

Appendix H
Modification of an Electromagnetic Wave by a Time-Varying Switched Magnetoplasma Medium: Transverse Propagation*

Joo Hwa Lee and Dikshitulu K. Kalluri

Abstract — The modification of frequency and field amplitudes of an extraordinary wave (*X* wave) by a time-varying magnetoplasma medium is considered. An explicit expression for the amplitude of the electric and magnetic fields in terms of the magnetoplasma parameters and the new frequency is obtained.

H.1 INTRODUCTION

The main effect of switching a magnetoplasma[1-17] is the splitting of the source wave into new waves whose frequencies are different from the frequency of the source wave. When the medium properties change slowly with time due to slow change of the electron density in the plasma, one of these waves whose initial frequency is the source frequency will be the dominant new wave in the sense that its field amplitudes will be significant. Other waves have field amplitudes of the order of the initial slope of the electron density profile. The case of longitudinal propagation has been already discussed in References 13, 16, and 17. The case of transverse propagation involving *X* waves[18] is the topic of this appendix.

* © 1998 IEEE. Reprinted with permission from Lee, J. H. and Kalluri, D. K., *IEEE Transactions on Plasma Science*, 26, 1–6, 1998.

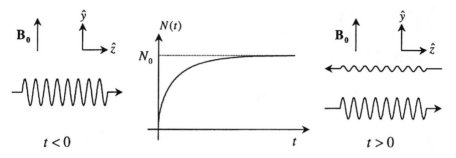

FIGURE H.1   Geometry of the problem. Modification of the source wave by a time-varying magnetoplasma medium: transverse propagation.

H.2 TRANSVERSE PROPAGATION

H.2.1 DEVELOPMENT OF THE PROBLEM

The geometry of the problem is shown in Figure H.1. Initially, for time $t < 0$, the entire space is considered to be free space. A uniform electromagnetic plane wave with a frequency ω_0 propagating in the positive z-direction and the electric field in the positive x-direction is established over the entire space.

The fields of the source wave are given by

$$\mathbf{E}(z,t) = \hat{x} E_0 e^{j(\omega_0 t - k_0 z)}, \qquad t < 0, \tag{H.1a}$$

$$\mathbf{H}(z,t) = \hat{y} H_0 e^{j(\omega_0 t - k_0 z)}, \qquad t < 0, \tag{H.1b}$$

where k_0 is the free-space wave number, $E_0 = \eta_0 H_0$, η_0 is the intrinsic impedance of free space, and $j = \sqrt{-1}$. A static magnetic field $\mathbf{B}_0$ is present along the y-direction.

At time $t = 0$, the entire space is converted into a time-varying plasma medium. The plasma medium is spatially homogeneous and unbounded. Therefore, the wave number k_0 remains constant.[1] The free-electron density $N(t)$ is slowly increasing to a certain value N_0 with zero initial value. Therefore, the plasma medium has a time-varying plasma frequency $\omega_p(t)$ and an electron gyrofrequency ω_b given by

$$\omega_p(t) = \sqrt{\frac{N(t)q}{m\varepsilon_0}}, \tag{H.2a}$$

$$\omega_b = \frac{B_0 q}{m}, \tag{H.2b}$$

where q is the absolute value of the charge, m the mass of the electron, and B_0 the magnitude of $\mathbf{B}_0$. With these assumptions, the fields for $t > 0$ can be written as follows:

$$\mathbf{E}(z,t) = \left[\hat{x}E_x(t) + \hat{z}E_z(t)\right]e^{-jk_0z}, \qquad t > 0, \tag{H.3a}$$

$$\mathbf{H}(z,t) = \hat{y}H_y(t)e^{-jk_0z}, \qquad\qquad t > 0, \tag{H.3b}$$

$$\mathbf{v}(z,t) = \left[\hat{x}v_x(t) + \hat{z}v_z(t)\right]e^{-jk_0z}, \qquad t > 0, \tag{H.3c}$$

where $\mathbf{v}(z,t)$ is the velocity of an electron. The differential equations for these fields can be obtained by using the Maxwell equations:

$$\nabla \times \mathbf{E} = -\mu_o \frac{\partial \mathbf{H}}{\partial t} \quad \text{and} \tag{H.4a}$$

$$\nabla \times \mathbf{H} = \varepsilon_0 \frac{\partial \mathbf{E}}{\partial t} + \mathbf{J}. \tag{H.4b}$$

From Equations H.3 and H.4, the following differential equations are obtained:

$$\eta_0 \frac{\partial H_y}{\partial t} = j\omega_0 E_x, \tag{H.5a}$$

$$\frac{\partial E_x}{\partial t} = j\omega_0 \eta_0 H_y - \frac{1}{\varepsilon_o} J_x, \tag{H.5b}$$

$$\frac{\partial E_z}{\partial t} = -\frac{1}{\varepsilon_0} J_z. \tag{H.5c}$$

Here, J_x and J_z denote the current density components. The current density vectors are closely related to the motion of the electron charges.

H.2.2 Current Density Vector

In the previous subsection, three equations with five unknowns were obtained. Two more equations are needed to determine the five unknowns. The equation of motion of the electrons is given by the Lorentz force equation

$$m\frac{d\mathbf{v}}{dt} = -q\left[\mathbf{E} + \mathbf{v} \times \mathbf{B}_0\right]. \tag{H.6}$$

The velocity vector may be related to the current density vector. When free space is slowly converted into plasma, it is necessary to impose the requirement that the electrons born at different times have zero initial velocity.[12,13] By applying these concepts to the case of transverse propagation,[14] the equations for J_x and J_z are obtained:

$$\frac{dJ_x}{dt} = \varepsilon_0 \omega_p^2 E_x + \omega_b J_z \quad \text{and} \tag{H.7a}$$

$$\frac{dJ_z}{dt} = \varepsilon_0 \omega_p^2 E_z - \omega_b J_x. \tag{H.7b}$$

H.2.3 HIGHER-ORDER DIFFERENTIAL EQUATION

Equations H.5 and H.7 are the first-order coupled differential equations that contain time-varying and complex coefficients. Generally, it is difficult to obtain exact solutions, but reasonable approximate solutions can be obtained. Equations H.5 and H.7 are manipulated by differentiation and substitution, and a fifth-order differential equation for E_x is obtained:

$$E_x^{(5)} + AE_x^{(3)} + BE_x^{(1)} + g\left[5E_x^{(2)} + \left(2\omega_o^2 + 3\omega_p^2\right)E_x\right] + \frac{dg}{dt}\left(3E_x^{(1)} + \omega_b E_z\right) + \frac{d^2g}{dt^2}E_x = 0, \tag{H.8}$$

where

$$A = \omega_0^2 + 2\omega_p^2 + \omega_b^2, \tag{H.9a}$$

$$B = \omega_0^2\omega_b^2 + \omega_0^2\omega_p^2 + \omega_p^4, \tag{H.9b}$$

$$g = \frac{d\omega_p^2}{dt}. \tag{H.9c}$$

dg/dt, d^2g/dt^2 and g^2, are neglected, since $\omega_p^2(t)$ is a slowly varying function of time. The equation for E_x becomes simpler and is given by

$$E_x^{(5)} + AE_x^{(3)} + BE_x^{(1)} + g\left[5E_x^{(2)} + \left(2\omega_0^2 + 3\omega_p^2\right)E_x\right] = 0. \tag{H.10}$$

H.2.4 RICCATI EQUATION

To solve Equation H.10, the Wentzel–Kramers–Brillouin (WKB) method can be applied.[15] A complex instantaneous frequency function, $p(t)$, is defined, such that

$$\frac{dE_x}{dt} = p(t)E_x. \tag{H.11}$$

Then,

$$E_x^{(2)} = (\dot{p} + p^2)E_x,$$ (H.12a)

$$E_x^{(3)} = (\ddot{p} + 3p\dot{p} + p^3)E_x,$$ (H.12b)

$$E_x^{(4)} = (\dddot{p} + 4p\ddot{p} + 6p^2\dot{p} + 3\dot{p}^2 + p^4)E_x,$$ (H.12c)

$$E_x^{(5)} = \left[\left(p^{(4)} + 5p\dddot{p} + 10(\dot{p} + p^2)\ddot{p} + 5(2p^3 + 3p\dot{p})\dot{p} + p^5\right)\right]E_x.$$ (H.12d)

Substitution of Equation H.12 into Equation H.10 gives

$$p^{(4)} + 5p\dddot{p} + 10(\dot{p} + p^2)\ddot{p} + 5(2p^3 + 3p\dot{p})\dot{p} + p^5$$
$$+ A(\ddot{p} + 3p\dot{p} + p^3) + Bp + g\left[5(\dot{p} + p^2) + (2\omega_0^2 + 3\omega_p^2)\right] = 0.$$ (H.13)

This Riccati equation can be solved to zero order, and the solution gives real
frequencies of the new waves.

H.2.5 FREQUENCIES OF THE NEW WAVES

If $p = j\omega$ in Equation H.13 and higher-order terms ($\dot{p}, \ddot{p}, \dddot{p}, p^{(4)}g$, and their powers)
are neglected, then

$$\omega^5 - A\omega^3 + B\omega = 0.$$ (H.14)

The solutions to this equation can be easily obtained:

For $\omega_b < \omega_0$

$$\omega_1 = \left[\tfrac{1}{2}\left\{A + \sqrt{A^2 - 4B}\right\}\right]^{1/2},$$ (H.15a)

$$\omega_2 = \left[\tfrac{1}{2}\left\{A - \sqrt{A^2 - 4B}\right\}\right]^{1/2},$$ (H.15b)

$$\omega_3 = -\omega_1,$$ (H.15c)

$$\omega_4 = -\omega_2,$$ (H.15d)

$$\omega_5 = 0;$$ (H.15e)

For $\omega_b > \omega_0$

$$\omega_1 = \left[\tfrac{1}{2}\left\{ A - \sqrt{A^2 - 4B} \right\} \right]^{1/2},$$

(H.16a)

$$\omega_2 = \left[\tfrac{1}{2}\left\{ A + \sqrt{A^2 - 4B} \right\} \right]^{1/2},$$

(H.16b)

$$\omega_3 = -\omega_1,$$

(H.16c)

$$\omega_4 = -\omega_2,$$

(H.16d)

$$\omega_5 = 0.$$

(H.16e)

The solutions are split into two regions given by Equations H.15 and H.16 to obtain the same initial values at $t = 0$. Consequently, in both regions, the first wave has the same initial value ω_0 and can be considered as the modified source wave (MSW). The other waves can be considered as new waves created by the switching action and are of the order of $g(0) = g_0$. Equations H.15 and H.16 are the same as those that can be obtained from the consideration of conservation of the wave number of the X wave in a time-varying, unbounded, and homogeneous plasma medium.[15]

Equations H.15 and H.16 show that five new waves are generated when $t > 0$ as a result of the imposition of the magnetoplasma. The first two solutions are transmitted waves, whereas the next two are reflected waves. Because ω_5 is zero, the fifth solution is not a wave.

H.2.6 AMPLITUDES OF THE FIELDS OF THE NEW WAVES

Amplitudes of the fields of the new waves can be obtained by expanding the frequency function to include a real part. Let $p = \alpha + j\omega$. Neglecting higher-order terms except α, $\dot{\omega}$, and g, the relation for α is obtained:

$$\alpha_n = \frac{g\left(5\omega_n^2 - 2\omega_0^2 - 3\omega_p^2\right) - \dot{\omega}\left(10\omega_n^2 - 3A\right)\omega_n}{5\omega_n^4 - 3A\omega_n^2 + B}.$$

(H.17)

H.2.7 COMPLETE SOLUTION

The complete solution to the problem can be represented by the linear combination of the five new modes. Suppose the field components have the following form:

$$E_x = E_0 \sum_{n=1}^{5} A_n \exp\left[\int_0^t \left(j\omega_n(\tau) + \alpha_n(\tau) \right) d\tau \right].$$

(H.18)

To determine the unknown coefficients A_n, the initial conditions at $t = 0$ can be used:

$$E_x(0) = E_0,$$ (H.19a)

$$E_z(0) = 0,$$ (H.19b)

$$H_y(0) = H_0,$$ (H.19c)

$$J_x(0) = 0,$$ (H.19d)

$$J_z(0) = 0.$$ (H.19e)

Symbolic manipulations[19] have been done to obtain the coefficients:

$$A_1 = 1 - jg_0 \frac{\omega_b^4 + 14\omega_b^2\omega_0^2 + \omega_0^4}{8\omega_0\left(\omega_0^2 - \omega_b^2\right)^3},$$ (H.20a)

$$A_2 = jg_0 \frac{\omega_b}{2\left(\omega_0 + \omega_b\right)\left(\omega_0 - \omega_b\right)^3},$$ (H.20b)

$$A_3 = jg_0 \frac{1}{8\omega_0\left(\omega_0^2 - \omega_b^2\right)},$$ (H.20c)

$$A_4 = -jg_0 \frac{\omega_b}{2\left(\omega_0 - \omega_b\right)\left(\omega_0 + \omega_b\right)^3},$$ (H.20d)

$$A_5 = 0.$$ (H.20e)

The above expressions become singular when $\omega_0 = \omega_b$. The evaluation of the unknown constants using Equation H.18 makes it possible to describe the transmitted fields and the reflected fields completely in terms of their time-varying envelope functions and the frequency functions. The resulting expression is

$$E_x(z,t) = E_0 \sum_{n=1}^{4} A_n \exp\left[\beta_n(t)\right] \exp\left[j\left(\theta_n(t) - k_0 z\right)\right],$$ (H.21)

where

$$\theta_n(t) = \int_0^t \omega_n(\tau)d\tau,$$ (H.22a)

$$\beta_n(t) = \int_0^t \alpha_n(\tau) d\tau. \tag{H.22b}$$

H.2.8 EXPLICIT EXPRESSIONS FOR THE AMPLITUDES OF THE FIELDS

The expression for g in terms of ω can be obtained by differentiating Equation H.14:

$$g = \frac{2(2\omega^2 - A)\omega\dot{\omega}}{2\omega^2 - \omega_0^2 - 2\omega_p^2}. \tag{H.23}$$

From Equations H.17 and H.23, $\alpha(t)$ can be expressed as

$$\alpha(t) = \frac{\left(\omega_0^2 A - 2\left(2\omega_b^2 + \omega_0^2\right)\omega^2\right)\dot{\omega}}{2\omega\left(2\omega^2\omega_b^2 - 3\omega_0^2\omega_b^2 + \omega_0^4 + 2\omega_b^2\omega_p^2\right)} = -\frac{2\omega\omega_b^2}{G(G - \omega_b^2)}\dot{\omega} - \frac{\omega_0^2}{2\omega G}\dot{\omega}, \tag{H.24}$$

where

$$G = \sqrt{4\omega^2\omega_b^2 - 4\omega_0^2\omega_b^2 + \omega_0^4} \tag{H.25a}$$

and

$$\omega_p^2 = \left(2\omega^2 - \omega_0^2 - G\right)/2. \tag{H.25b}$$

Equation H.24 can be integrated to obtain $\beta(t)$:

$$\beta(t) = \int_0^t \alpha(\tau) d\tau = -\frac{1}{2}\ln\left(G - \omega_b^2\right) + \frac{\omega_b^2}{2\sqrt{\omega_0^4 - 4\omega_0^2\omega_b^2}}\ln\frac{G + \sqrt{\omega_0^4 - 4\omega_0^2\omega_b^2}}{\omega}\Bigg|_{\omega(0)}^{\omega(t)}. \tag{H.26}$$

H.2.9 ADIABATIC ANALYSIS OF THE MODIFIED SOURCE WAVE

For the first wave, from Equation H.26,

$$F(t) = \exp[\beta_1(t)] = \sqrt{\frac{\omega_0^2 - \omega_b^2}{2\omega_1^2 - \omega_0^2 - 2\omega_p^2 - \omega_b^2}}$$

$$\times \left(\frac{\omega_0}{\omega_1}\frac{2\omega_1^2 - \omega_0^2 - 2\omega_p^2 + \sqrt{\omega_b^4 - 4\omega_b^2\omega_0^2}}{\omega_0^2 + \sqrt{\omega_0^4 - 4\omega_b^2\omega_0^2}}\right)^{\frac{\omega_b^2}{2\sqrt{\omega_0^4 - 4\omega_b^2\omega_0^2}}}, \tag{H.27}$$

In Equation H.27, the factors, including the last exponent, are real for $\omega_0^2 > 4\omega_b^2$. For other ranges, some of these factors are complex but easily computed by using mathematical software.[15] For small g_0, and ω_0 sufficiently different from ω_b, $A_1 \cong 1$ and the other A coefficients are negligible. Only the first wave has significant amplitude, and it can be labeled as the MSW. Each of the fields of the MSW can be written as a product of an instantaneous amplitude function and a frequency function

$$E_x(z,t) = E_0 e_x(t)\cos\left[\theta_1(t) - k_0 z\right],$$ (H.28a)

$$H_y(z,t) = H_0 h_y(t)\cos\left[\theta_1(t) - k_0 z\right],$$ (H.28b)

$$E_z(z,t) = E_0 e_z(t)\cos\left[\theta_1(t) - k_0 z + \frac{\pi}{2}\right],$$ (H.28c)

where the instantaneous amplitude functions are given by

$$e_x(t) = F(t),$$ (H.28d)

$$h_y(t) = \frac{\omega_0}{\omega_1(t)} F(t),$$ (H.28e)

$$e_z(t) = \frac{\omega_b \omega_p^2}{\omega_1(t)\left[\omega_1^2(t) - \omega_p^2(t) - \omega_b^2\right]} F(t),$$ (H.28f)

and the power density function $S(t)$ is given by

$$S(t) = \frac{\omega_0}{\omega_1(t)}\left[F(t)\right]^2.$$ (H.29)

For $\omega_b \ll \omega_0$ the amplitude functions reduce to

$$e_x(t) \cong \left(\frac{\omega_0^2}{\omega_0^2 + \omega_p^2(t)}\right)^{1/4}\left[1 + \frac{\left(2\omega_0^2\left(\omega_0^2 + \omega_p^2(t)\right)\ln\left(\frac{\omega_0^2}{\omega_0^2 + \omega_p^2(t)}\right) - \omega_p^2(t)\left(3\omega_0^2 + 2\omega_p^2(t)\right)\right)\omega_b^2}{4\omega_0^4\left(\omega_0^2 + \omega_p^2(t)\right)}\right],$$ (H.30a)

$$e_z(t) \cong \left(\frac{\omega_0^2}{\omega_0^2 + \omega_p^2(t)}\right)^{1/4}\frac{\omega_b \omega_p^2(t)}{\omega_0^2\sqrt{\omega_0^2 + \omega_p^2(t)}}.$$ (H.30b)

For $\omega_b \gg \omega_0$ and $\omega_b \gg \omega_p$,

$$e_x(t) \cong 1 - \frac{3}{4}\frac{\omega_p^2(t)}{\omega_b^2} \quad \text{and} \tag{H.30c}$$

$$e_z(t) \cong -\frac{\omega_p^2(t)}{\omega_b\omega_0}. \tag{H.30d}$$

Graphical results given in Section H.3 are obtained by computing $F(t)$ from Equation H.27. The results are verified using Equation H.22b involving a numerical integration routine.

H.3 GRAPHICAL ILLUSTRATIONS AND RESULTS

An exponentially varying electron density profile is used for an illustrative example. The time-varying plasma frequency is considered as:

$$\omega_p^2(t) = \omega_{p0}^2\left[1 - \exp(-bt)\right]. \tag{H.31}$$

The results for the first wave are presented in a normalized form by taking ω_0 = 1. The parameter b is chosen as 0.01. Figures H.2 through H.4 show the characteristics of the MSW. It has an upshifted frequency (Figure H.2) compared with the source wave frequency ω_0 when $\omega_b < \omega_0$ and a downshifted frequency when $\omega_b > \omega_0$. The amplitude function $e_x(t)$ (Figure H.3) of the MSW decreases with the increase of ω_b as long as $\omega_b < \omega_0$. However, it increases with ω_b for $\omega_b > \omega_0$. The amplitude function $h_y(t)$ follows a similar pattern. The amplitude function $e_z(t)$ changes sign as the ω_b value crosses $\omega_b = \omega_0$ (Figure H.4).

Figures H.5 through H.8 show the variation of the saturation values ($t \to \infty$) of the frequency and amplitude function of the MSW. The horizontal axis is ω_b. However, in the neighborhood of $\omega_b \approx \omega_0$, the assumptions that $A_1 \approx 1$ and all other A values are zero are not valid. The following validity condition is obtained by restricting the magnitude of the second term on the right-hand side of Equation H.20a to be less than 0.1. The validity condition is

$$\left|1 - \frac{\omega_b^2}{\omega_0^2}\right| > \left(20\frac{b\omega_{p0}^2}{\omega_0^3}\right)^{1/3}. \tag{H.32}$$

From Figures H.5 through H.8, the following qualitative result may be stated: as the value of ω_b crosses ω_0, there will be a rapid change in the frequency and the fields of the source wave.

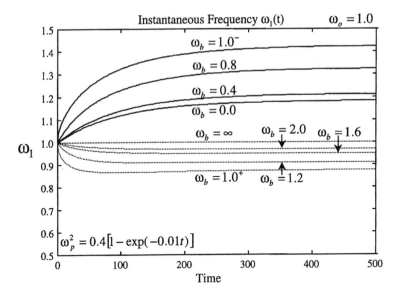

FIGURE H.2   Instantaneous frequency of the MSW.

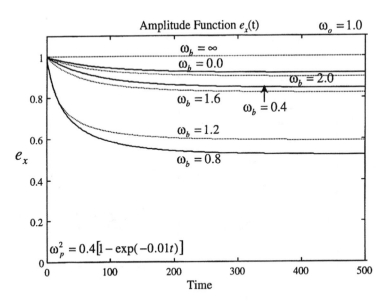

FIGURE H.3   Amplitude function of the x component of the electric field of MSW.

H.4 CONCLUSION

The creation of a time-varying magnetoplasma medium splits the incident wave into four new waves having different frequencies in the case of transverse propagation. Two of the waves are transmitted waves, and the other two are reflected waves. The

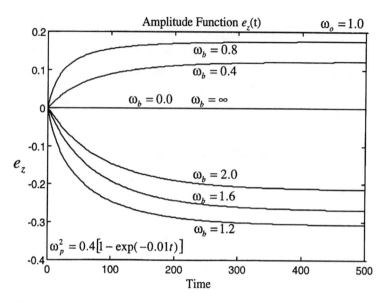

FIGURE H.4 Amplitude function of the z component of the electric field of MSW.

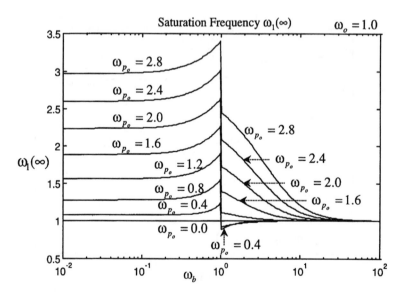

FIGURE H.5 Variation of the saturation frequency of the MSW with the electron cyclotron frequency (ω_b) and the saturation plasma frequency (ω_{p0}).

first wave whose initial frequency is that of the incident wave has the strongest intensity compared with the others and is labeled the MSW. Based on the adiabatic analysis of the MSW, the following qualitative statement can be made: as the value of ω_b crosses ω_0, there will be a rapid change in the frequency and the fields.

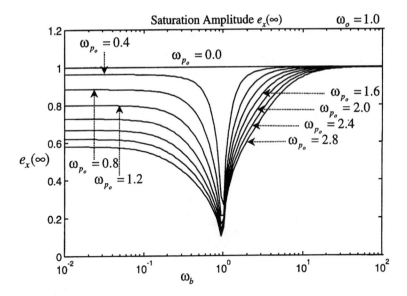

FIGURE H.6 Variation of the saturation amplitude of the x component of the electric field of the MSW with ω_b and ω_{p0}.

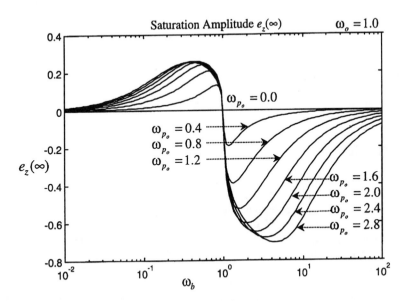

FIGURE H.7 Variation of the saturation amplitude of the z component of the electric field of the MSW with ω_b and ω_{p0}.

There is still difficulty quantifying the behavior of the fields in the neighborhood of $\omega_b \approx \omega_0$. The accuracy of the solution degrades since $\dot{\omega}$ at $t = 0$ tends to infinity as ω_b tends toward ω_0. In this neighborhood, the higher-order terms become important.

Saturation Power density $S(\infty)$

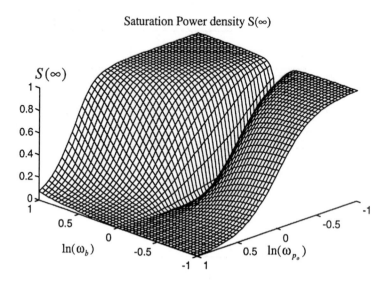

FIGURE H.8 Variation of the saturation power density of the MSW with ω_b and ω_{p0}.

The aspect of the effect at $\omega_b = \omega_0$ is currently under investigation using a very different approach, and the result will be reported in due course.

Adiabatic solutions given in this appendix can give explicit analytical description for the frequency and the amplitude of the modified source wave except in the neighborhood of $\omega_b \approx \omega_0$.

ACKNOWLEDGMENT

The authors are thankful to the referees for their valuable comments.

REFERENCES

1. Jiang, C. L., Wave propagation and dipole radiation in a suddenly created plasma, *IEEE Trans. Antennas Propag.*, AP-23, 83, 1975.
2. Kalluri, D. K., On reflection from a suddenly created plasma half-space: transient solution, *IEEE Trans. Plasma Sci.*, 16, 11, 1988.
3. Goteti, V. R. and Kalluri, D. K., Wave propagation in a switched-on time-varying plasma medium, *IEEE Trans. Plasma Sci.*, 17, 828, 1989.
4. Kalluri, D. K. and Goteti, V. R., Frequency shifting of electromagnetic radiation by sudden creation of a plasma slab, *J. Appl. Phys.*, 72, 4575, 1992.
5. Joshi, C. J., Clayton, C. E., Marsh, K., Hopkins, D. B., Sessler, A., and Whittum, D., Demonstration of the frequency upshifting of microwave radiation by rapid plasma creation, *IEEE Trans. Plasma Sci.*, 18, 814, 1990.
6. Kuo, S. P. and Ren, A., Experimental study of wave propagation through a rapidly created plasma, *IEEE Trans. Plasma Sci.*, 21, 53, 1993.
7. Kalluri, D. K. and Goteti, V. R., Damping rates of waves in a switched magnetoplasma medium: longitudinal propagation, *IEEE Trans. Plasma Sci.*, 18, 797, 1993.

8. Goteti, V. R. and Kalluri, D. K., Wave propagation in a switched magnetoplasma medium: transverse propagation, *Radio Sci.*, 25, 61, 1990.

9. Madala, S. R. V. and Kalluri, D. K., Frequency shifting of a wave traveling in an isotropic plasma medium due to slow switching of a magnetic field, *Plasma Sourc. Sci. Technol.*, 1, 242, 1992.

10. Madala, S. R. V. and Kalluri, D. K., Longitudinal propagation of low-frequency waves in a switched magnetoplasma medium, *Radio Sci.*, 28, 121, 1993.

11. Lai, C. H. and Katsouleas, T. C., Frequency upshifting by an ionization front in a magnetized plasma, *IEEE Trans. Plasma Sci.*, 21, 45, 1993.

12. Baños, A., Jr., Mori, W. B., and Dawson, J. M., Computation of the electric and magnetic fields induced in a plasma created by ionization lasting a finite interval of time, *IEEE Trans. Plasma Sci.*, 21, 57, 1993

13. Kalluri, D. K., Goteti, V. R., and Sessler, A. M., WKB solution for wave propagation in a time-varying magnetoplasma medium: longitudinal propagation, *IEEE Trans. Plasma Sci.*, 21, 70, 1993.

14. Lee, J. H., Wave Propagation in a Time-Varying Switched Magnetoplasma Medium: Transverse Propagation, M.S. Thesis, University of Massachusetts Lowell, Lowell, 1994.

15. Kalluri, D. K., Frequency shifting using magnetoplasma medium: flash ionization, *IEEE Trans. Plasma Sci.*, 21, 77, 1993.

16. Kalluri, D. K., Conversion of a whistler wave into a controllable helical wiggler magnetic field, *J. Appl. Phys.*, 79, 6770, 1996.

17. Kalluri, D. K., Frequency upshifting with power intensification of a whistler wave by a collapsing plasma medium, *J. Appl. Phys.*, 79, 3895, 1996.

18. Heald, M. A. and Wharton, C. B., *Plasma Diagnostics with Microwaves*, Wiley, New York, 1965, 36.

19. Blachman, N., *Mathematica: A Practical Approach*, Prentice-Hall, Englewood Cliffs, NJ, 1992.

Problems

P1.1. For a lossy plasma, the frequency domain relative permittivity is complex and is given by

$$\varepsilon_p(\omega) = 1 - \frac{\omega_p^2}{\omega(\omega - j\nu)}, \tag{P1.1}$$

where ν is the collision frequency (rad/s). The above relation is called the Drude dispersion relation and in the limit $\nu \to 0$ reduces to Equation 1.50.

Show that the basic field equations for a cold isotropic lossy plasma are given by Equations 1.1, 1.2, and

$$\frac{d\mathbf{J}}{dt} + \nu \mathbf{J} = \varepsilon_0 \omega_p^2 \mathbf{E}. \tag{P1.2}$$

Consider a plasma of plasma frequency $f_p = \omega_p/2\pi = 30$ GHz and a collision frequency $\nu = 2 \times 10^{10}$ rad/s. Plot the real and imaginary parts of the dielectric constant in the range of 0 to 90 GHz.

P1.2. Water is an example of materials whose frequency domain relative permittivity exhibits Debye dispersion:

$$\varepsilon_r(\omega) = \varepsilon_\infty + \frac{\varepsilon_s - \varepsilon_\infty}{1 + j\omega t_0}, \tag{P1.3}$$

where ε_∞ is the infinite frequency relative permittivity, ε_s is the static relative permittivity at zero frequency, and t_0 is the relaxation time.

Show that the basic field equations in this medium may be described by

$$\nabla \times \mathbf{E} = -\mu_0 \frac{\partial \mathbf{H}}{\partial t}, \tag{P1.4}$$

$$\nabla \times \mathbf{H} = \frac{\partial \mathbf{D}}{\partial t}, \tag{P1.5}$$

$$t_0 \frac{d\mathbf{D}}{dt} + \mathbf{D} = \varepsilon_s \varepsilon_0 \mathbf{E} + t_0 \varepsilon_\infty \varepsilon_0 \frac{d\mathbf{E}}{dt}, \tag{P1.6}$$

and the one-dimensional field equation (Equations 1.20 through 1.22) are modified as

$$\frac{\partial E}{\partial z} = -\mu_0 \frac{\partial H}{\partial t}, \tag{P1.7}$$

$$-\frac{\partial H}{\partial z} = \frac{\partial D}{\partial t}, \tag{P1.8}$$

$$t_0 \frac{dD}{dt} + D = \varepsilon_s \varepsilon_0 E + t_0 \varepsilon_\infty \varepsilon_0 \frac{dE}{dt}. \tag{P1.9}$$

Plot the real and imaginary parts of complex relative permittivity of water in the frequency domain 0 to 80 GHz. Assume $\varepsilon_s = 81$, $\varepsilon_\infty = 1.8$, and $t_0 = 9.4 \times 10^{-12}$ s.

P1.3. A second-order Lorentz dispersive material has a relative permittivity

$$\varepsilon_r(\omega) = \varepsilon_\infty + \frac{(\varepsilon_s - \varepsilon_\infty)\omega_R^2}{\omega_R^2 + 2j\omega\delta - \omega^2}, \tag{P1.10}$$

where ω_R is the resonant frequency and δ is the damping constant.
Show that the basic field equations in this medium are given by

$$\nabla \times \mathbf{E} = -\mu_0 \frac{\partial \mathbf{H}}{\partial t}, \tag{P1.11}$$

$$\nabla \times \mathbf{H} = \frac{\partial \mathbf{D}}{\partial t}, \tag{P1.12}$$

$$\omega_R^2 \mathbf{D} + 2\delta \frac{d\mathbf{D}}{dt} + \frac{d^2\mathbf{D}}{dt^2} = \omega_R^2 \varepsilon_s \varepsilon_0 \mathbf{E} + 2\delta \varepsilon_\infty \varepsilon_0 \frac{d\mathbf{E}}{dt} + \varepsilon_\infty \varepsilon_0 \frac{d^2\mathbf{E}}{dt^2}, \tag{P1.13}$$

and the one-dimensional field equations (Equations 1.20 through 1.22) are modified as

$$\frac{\partial E}{\partial z} = -\mu_0 \frac{\partial D}{\partial t}, \tag{P1.14}$$

$$-\frac{\partial H}{\partial z} = \frac{\partial D}{\partial t}, \tag{P1.15}$$

$$\omega_R^2 D + 2\delta \frac{dD}{dt} + \mu_0 \frac{d^2 D}{dt^2} = \omega_R^2 \varepsilon_s \varepsilon_0 E + 2\delta \varepsilon_\infty \varepsilon_0 \frac{dE}{dt} + \varepsilon_\infty \varepsilon_0 \frac{d^2 E}{dt^2}. \qquad (P1.16)$$

Plot the real and imaginary parts of $\varepsilon_r(\omega)$ using the parameters $\varepsilon_s = 2.25$, $\varepsilon_\infty = 1$, $\omega_R = 4 \times 10^{16}$ rad/s, $\delta = 0.28 \times 10^{16}$/s. This Lorentz medium has a resonance in the optical range.

CHAPTER 2

P2.1. Let $\varepsilon_p(z,\omega)$ in Equation 2.15 be given by

$$\varepsilon_p(z,\omega) = \varepsilon_1, \qquad\qquad -\infty < z < 0, \qquad (P2.1)$$

$$= \varepsilon_1 + (\varepsilon_2 - \varepsilon_1)\frac{z}{L}, \quad 0 < z < L, \qquad (P2.2)$$

$$= \varepsilon_2, \qquad\qquad L < z < \infty. \qquad (P2.3)$$

Show that the solution of Equation 2.15 in the two regions given below can be written as

$$E = Ie^{-jk_1 z} + Re^{+jk_1 z}, \qquad -\infty < z < 0, \qquad (P2.4)$$

$$= Te^{-jk_2 z} + Ae^{jk_2 z}, \qquad L < z < \infty, \qquad (P2.5)$$

where I, R, T, and A are constants and $k_1 = k_0\sqrt{\varepsilon_1}$, $k_2 = k_0\sqrt{\varepsilon_2}$. Give physical reasons for setting $A = 0$. Note that if $I = 1$, R and T give the reflection and transmission coefficients. Assume the waves are propagating along the z-axis.

P2.2. In this problem the nature of the solution of Equation 2.15 in the domain $0 < z < L$ is investigated.

By introducing a new variable $\xi = k_0^2 \varepsilon_p(z,\omega)$, show that Equation 2.15 may be transformed to

$$\frac{d^2 E}{d\xi^2} + \frac{\xi E}{\beta^2} = 0, \qquad (P2.6)$$

where

$$\beta = k_0^2 \frac{\varepsilon_2 - \varepsilon_1}{L}. \qquad (P2.7)$$

Note that the solution of Equation P2.6 can be written in terms of Airy function Ai and Bi

$$E = c_1 Ai\left[-\beta^{-2/3}\xi\right] + c_2 Bi\left[-\beta^{2/3}\xi\right], \qquad 0 < z < L. \tag{P2.8}$$

P2.3. Determine the reflection coefficient R and plot the power reflection coefficient $|R|^2$ vs. L. Take $\varepsilon_1 = (4/3)^2$ and $\varepsilon_2 = 1$. These data approximate the water–air interface at optical frequencies. Assume the dielectric function is a linear profile given in Equations P2.1 through P2.3.

P2.4. Find an analytical expression for $|R|^2$ if $L = 0$ in Equations P2.3 and P2.2.

P2.5. Plot $|R|^2$ vs. θ_i for the data of Equation P2.4 if θ_i is the angle of incidence.

P2.6. Let $z < 0$ be free space and $z > 0$ be a dispersive medium. For the data given in Problem P1.1, find ρ and τ. Find the absorptance defined by $A = 1 - \rho - \tau$. Justify the definition. Plot A vs. f, ρ vs. f, and τ vs. f in the frequency band $0 < f < 90$ GHz. Assume normal incidence.

P2.7. Let $z > 0$ be water and $z < 0$ free space. Using the data of Problem P1.2, plot A vs. f, ρ vs. f, and τ vs. f. Assume normal incidence.

P2.8. Let $z > 0$ be the second-order Lorentz dispersive medium and $z < 0$ free space. Using the data of Problem P1.3, plot A vs. ω, ρ vs. ω, and τ vs. ω in the frequency band $2 \times 10^{16} < \omega < 6 \times 10^{16}$. Comment on the values of ρ, τ, and A around the frequency $\omega \approx 4 \times 10^{16}$/s. Repeat the above calculations for $\xi = 1.4 \times 10^{16}$/s. What do you infer from the two sets of results? Assume normal incidence.

P2.9. Discuss the possibility of having a TE surface wave mode at a metal–vacuum interface.

P2.10. Light of wavelength 0.633 μm is incident from air on to an air–aluminum interface. Take the dielectric constant of aluminum at 0.633 μm as $\varepsilon_r = -60.56 - j24.86$.

Plot reflectivity (ρ) vs. θ_i of an s wave, where θ_i is the angle of incidence.

P2.11. The half-space $z > 0$ is free space and the half-space $z < 0$ is metal of plasma frequency $\omega_p = 8.2 \times 10^{15}$ rad/s. Find the reflectivity of a p wave in free space incident on the metal at an angle of incidence of 45°, when

a. $\omega = 8.2 \times 10^{15}$ rad/s

b. $\omega = \sqrt{2}\ (1.1)8.2 \times 10^{15}$ rad/s

P2.12. The interface $z = 0$ is as described in Problem P2.11. Investigate the propagation of a surface plasmon (TM) on this interface at $\omega = 4 \times 10^{15}$ rad/s.

a. Determine the phase velocity of the surface wave on the interface.

b. Determine the distance, in meters, in which the z-component of the electric field in free space reaches 0.3679 ($=1/e$).

c. Determine the distance, in meters, in which the z-component of the electric field in metal reaches 0.3679 ($= 1/e$) of its value at the interface.

P2.13. The interface $z = 0$ is as described in Problem P2.11. Investigate the propagation of a surface plasmon (TM) on this interface at $\omega = 9.84 \times 10^{15}$ rad/s.

a. Determine the phase velocity of the surface wave on the interface.

b. Determine the distance, in meters, in which the z-component of the electric field in free space reaches $0.3679 \, (= 1/e)$ of its value at the interface.

c. Determine the distance, in meters, in which the z-component of the electric field in metal reaches $0.3679 \, (= 1/e)$ of its value at the interface.

P2.14. Derive the dispersion relation and draw the ω–β diagram of an unbounded periodic media consisting of alternating layers of dielectric layers. Assume the following values for the parameters of the layers in a unit cell:

First layer: $\varepsilon_1 = 2\varepsilon_0$, $\mu_1 = \mu_0$, $L_1 = L_0 = 3$ cm
Second layer: $\varepsilon_2 = 3\varepsilon_0$, $\mu_2 = \mu_0$, $L_2 = 2L_0$

Determine the location of the center of the first stop band. Also, determine the bandwidth of this stop band.

CHAPTER 3

P3.1. a. Equations 3.25 through 3.27 specify the initial conditions for the suddenly created plasma. Convert these initial conditions to the initial conditions on H, $\dot{H}$, and $\ddot{H}$

b. Solve Equation 3.13 with ω_p^2 constant and show that Equation 3.15 will be the solution of the higher-order differential equation in H. Determine ω_m and H_m in Equation 3.15.

P3.2. Let ω_1 be the frequency of a propagating wave in a plasma medium of plasma frequency ω_{p1}. At $t = 0$, the electron density suddenly drops. Let $\omega_p = \omega_{p2}$ for $t > 0$, where $\omega_{p2} < \omega_{p1}$. Find the new frequencies and fields of all the modes generated by the switching action.

P3.3. Let the plasma frequency be a step profile in space, i.e.,

$$\omega_p^2(z,t) = \omega_{p1}^2, \quad z < 0$$

$$\omega_p^2(z,t) = \omega_{p2}^2, \quad z > 0$$

a. Find the Green's function for this problem of spatially unlike media.

b. Develop a perturbation technique for this problem on the lines of Section 3.5, where $\omega_p^2(z)$ is a fast profile with a scale length L.

c. Consider a linear profile with a slope of $(\omega_{p2}^2 - \omega_{p1}^2)/L$. Determine the reflection coefficient for this profile to the order of L^2, i.e., you may neglect terms in L^3 and higher powers of L. Assume that the source wave has a frequency ω_1 and wave number k_1.

P3.4. Solve the problem specified in Problem P3.3 using the exact solution on the lines of Section 3.9 and express the solution in terms of Airy functions. Use a mathematical software to plot the power reflection coefficient ρ vs. the scale length L. Compare this result with the result obtained in Problem 3.3c.

Chapter 4

P4.1. a. Show that the Laplace transform of the electric field of the reflected field when the suddenly created plasma is lossy with a collision frequency ν is given by

$$\frac{E_{yr}(0,s)}{E_0} = \frac{N_1 N_2}{D_1 D_2}, \tag{P4.1}$$

$$N_1(s) = \omega_0^2 A - s^2 \sqrt{A^2 + \omega_p^2}, \tag{P4.2}$$

$$N_2(s) = \sqrt{A^2 + \omega_p^2} - A, \tag{P4.3}$$

$$D_1(s) = s^2 + \omega_0^2, \tag{P4.4}$$

$$D_2(s) = s^3 + \nu s^2 + s\left(s^2 + \omega_p^2\right) + \nu \omega_0^2, \tag{P4.5}$$

$$A(s) = \sqrt{s(s+\nu)}. \tag{P4.6}$$

b. For the case of a low-loss plasma ($\nu/\omega_p \ll 1$), show that $D_2(s)$ may be factored as

$$D_2(s) = \left[s + \frac{\nu \omega_0^2}{\omega^2}\right]\left[(s+\alpha_1)^2 + \omega^2\right], \tag{P4.7}$$

where

$$\alpha_1 = \frac{\nu}{2} \frac{\omega_p^2}{\omega^2}, \tag{P4.8}$$

$$\omega^2 = \omega_0^2 + \omega_p^2. \tag{P4.9}$$

c. From the results of (b) deduce that the damping time constant t_{pB} of the B wave is

$$t_{pB} = \frac{2\omega^2}{\nu\omega_p^2},$$
(P4.10)

and the damping time constant of the wiggler magnetic field is

$$t_w = \frac{\omega^2}{\nu\omega_0^2}.$$
(P4.11)

P4.2. Let $f_0 = 10$ GHz. Find N_0 if the frequency of the upshifted wave is 23 GHz. If $\nu/\omega_p = 0.01$, find the collision frequency. Calculate the damping constants t_p and z_p.

P4.3. For the data given in Problem P4.2 find the value of z_1 in Figure 4.2 if the negative-going B wave attenuates to 50% of the original value at z_1 by the time it reaches the interface.

CHAPTER 5

P5.1. Show that the sum of $R_A(s)$ and $R_B(s)$ given by Equations 5.32 and 5.33 equals $A_{1R}(s)/E_0$ given by Equation 4.42.

P5.2. Show that in the limit $d \to \infty$, Equation 5.24 becomes Equation 5.32 and Equation 5.25 becomes Equation 5.33.

P5.3. For the data given in Problem P4.2, find the minimum slab width d, if the strength of the A wave, in the region $z > d$, is negligible (less than 1% of the strength of the incident wave).

CHAPTER 6

P6.1. Derive Equation 6.25.

P6.2. Show that Equations 6.25 and 6.28 are the same.

P6.3. Given $f = 2.8$ GHz and $B_0 = 0.3T$ find the electron density of the plasma at which the L wave has a cutoff.

P6.4. Faraday rotation of an 8-mm wavelength microwave beam in a uniform plasma in a $0.1T$ magnetic field is measured. The plane of polarization is found to be rotated by 90° after traversing 1 m of plasma. What is the electron density in the plasma?

P6.5. Derive Equations 6.52 and 6.54.

P6.6. Derive Equations 6.56, 6.57, and 6.58.

P6.7. a. For a whistler mode, i.e., $\omega \ll \omega_p$, $\omega \ll \omega_b$, show that the group velocity is given approximately by

$$v_g \sim \frac{\sqrt{\omega\omega_b}}{\omega_p}.$$
(P6.1)

b. Assume that, as a result of lightning, signals in the frequency range of 1 to 10 kHz are generated. If these signals are guided from one hemisphere to the other hemisphere of the Earth along Earth's magnetic field over a path length of 5000 km, calculate the travel time of the signals as a function of frequency. Take the following representative values for the parameters: $f_p = 0.5$ MHz and $f_b = 1.5$ MHz.

P6.8. A right circularly polarized electromagnetic wave of frequency $f_0 = 60$ Hz is propagating in the z-direction in the metal potassium, at a very low temperature of a few degrees Kelvin, in the presence of a z-directed static magnetic flux density field B_0.

Determine (a) the phase velocity and (b) the wavelength of the electromagnetic wave in the metal.

Assume that the electron density N_0 of the metal is such that $\omega_p = 10^{16}$ rad/s

Assume that the B_0 value is such that the electron gyrofrequency $\omega_b = 10^{12}$ rad/s

Assume that the collision frequency v is negligible

The calculations become simple if approximations are made based on the inequalities mentioned above. Relate the specification of a very low temperature to one of the assumptions given.

P6.9. A right circularly polarized wave is propagating in the z-direction in a semiconductor in the presence of a z-directed B_0 field. Assuming $\omega_b \gg \omega_p$ and $v/\omega_b \ll 1$, investigate the wave propagation near cyclotron resonance $\omega = \omega_b$; i.e., find expression for $k(\omega)$, where k is a complex wave number.

Plot Im $k(\omega)$/Im $k(\omega_b)$ vs. ω/ω_b for two values of the parameter, (a) $v/\omega_b = 0.1$ and (b) $v/\omega_b = 0.01$. Assume $\omega_b = 10^{12}$ rad/s and $\omega_p = 10^{10}$ rad/s for computation.

Discuss the resonance characteristic of the response indicated by the results of (a) and (b). Suppose this sample is put in a waveguide and an electromagnetic wave of frequency ω (R wave) is launched. Then a B_0 field in z-direction is applied and the output of the electromagnetic wave measured, while the strength of B_0 is varied. For what value of ω will the output be minimal?

P6.10. An electromagnetic wave of frequency ω and wave number k is propagating in a magnetoplasma medium at an angle θ to the direction of the static magnetic field. Determine the dispersion relation and sketch the ω–k diagram. Identify the cutoff frequencies and the resonant frequencies.

P6.11. A p wave in free space of frequency ω_0 is obliquely incident on a magnetoplasma half-space. Let $z < 0$ be free space and $z > 0$ be the magnetoplasma half-space and let the static magnetic field be in the z-direction. Let the angle of incidence with the z-axis be θ_I. Find the power reflection coefficient ρ. Express your answer in terms of θ_I, ω_b, ω_p, and ω_0. Examine the existence of the Brewster angle for this problem.

P6.12. Study the dispersion relation of a surface magnetoplasmon.

CHAPTER 7

P7.1. Show that the wave impedance given by Equation 7.20 is the same as that given by Equation 7.30.

P7.2. The source frequency f_0 in free space is 10 GHz. The frequency upshift ratio of the R1 wave is 1.5. If $\omega_b = \omega_p$, find N_0 and $\mathbf{B}_0$. Find the damping constants t_p and z_p if $v/\omega_p = 0.01$.

P7.3. Repeat the calculation of Problem P7.2 for L1, X1, and X2 waves (see Appendix B for the R and L waves and Appendix C for the X waves).

P7.4. Let $\omega_b = 0$. Repeat the calculation of Problem P7.2 for the O wave.

P7.5. The frequency downshift ratio is given as 0.4. Repeat the calculations of Problem P7.2 for the R3 wave.

P7.6. The frequency downshift ratio is given as 0.7. Repeat the calculations of Problem P7.2 for the X2 wave (see Appendix C for the X wave).

P7.7. A perpendicularly polarized wave (s wave) of frequency ω_0 and wave number k is propagating in free space in the presence of a static magnetic field $\mathbf{B}_0$ which is in the z-direction. The angle between $\mathbf{B}_0$ and the wave vector $\mathbf{k}$ is θ_I. At $t = 0$ an unbounded plasma medium of plasma frequency ω_p is created. Find the frequencies and the fields of the newly created wave modes. Assume that the plasma is lossless.

CHAPTER 8

P8.1. Describe the differential equation for longitudinal propagation of an R wave in an inhomogeneous magnetoplasma with a space-varying profile $\omega_p^2(z)$. The gyrofrequency ω_b is constant.

P8.2. Find the reflection and transmission coefficients if $\omega_p^2(z)$ is a spatial step profile $\tilde{\omega}_p^2(z)$. The step change is from ω_{p1} to ω_{p2} at $z = 0$.

P8.3. Find the Green's function for longitudinal propagation of an R wave in spatially unlike media. Medium 1 has a plasma frequency ω_{p1} and medium 2 has a plasma frequency ω_{p2}. The gyrofrequency is ω_b.

P8.4. Consider a linear profile for the electron density, i.e.,

$$\omega_p^2(z) = \begin{cases} 0 & z < -L/2, \\ \dfrac{\omega_p^2}{2} + \dfrac{\omega_p^2}{L}z, & -L/2 < z < L/2, \\ \omega_p^2, & z > L/2. \end{cases} \tag{P8.1}$$

Calculate the power reflection coefficient of a source wave of frequency ω_1 and wave number k_1. Assume that the profile is a fast profile so that terms in L^3 and higher may be neglected.

CHAPTER 9

P9.1. Perform the adiabatic analysis of longitudinal propagation of an R wave in a magnetized plasma with spatially varying electron density profile $\omega_p^2(z)$. Find $E_m(z)/E_0$ when $E_m(z)$ is the amplitude of the modified source wave and E_0 is the amplitude at $z = 0$.

P9.2. An R wave of frequency ω_0 is propagating in anisotropic plasma of plasma frequency ω_p. At $t = 0$, a slowly varying nonoscillatory magnetic field in the direction of propagation is created, i.e.,

$$\omega_b = 0, \quad t < 0, \tag{P9.1}$$

$$\omega_b = \omega_b(t), \ t > 0, \tag{P9.2}$$

where ω_b varies slowly with time. Apply the adiabatic analysis to determine the modification of the frequency and the magnetic field of the source wave.

CHAPTER 10

P10.1. An O wave of frequency ω_0 is propagating in free space. At $t = 0$, a Lorentz medium discussed in Section 10.5 is suddenly created. Determine the frequencies and the fields of the frequency-shifted waves created by the switching action.

P10.2. Consider that a Lorentz medium is created suddenly in the presence of a static magnetic field B_0 in the direction of propagation of an R source wave of frequency ω_0. Determine the frequency and the fields of the frequency-shifted waves created by the switching action.

P10.3. A source wave of frequency of 10 GHz is propagating in free space. At $t = 0$, an unbounded periodic media of plasma layers shown in Figure 2.1 is created. If $L = 1.8$ cm, $d = 0.6$ cm, and $f_{p0} = 12$ GHz, determine the frequencies of the first three modes in the increasing order of frequencies (positive numbers). Identify the downshifted frequency mode.

P10.4. Show that R and L waves are the natural modes of propagation in a chiral medium. Find the wave number and the wave impedance of (a) the R wave and (b) the L wave.

P10.5. a. Draw the ω–k diagram for an unbounded periodic media consisting of alternating layers of chiral media. Assume the following values for the parameters of the layers in a unit cell
First layer: $\varepsilon_1 = 2\varepsilon_0$, $\mu_1 = \mu_0$, $\xi_{c1} = 10^{-3}$, $L_1 = L_0 = 3$ cm
Second layer: $\varepsilon_2 = 3\varepsilon_0$, $\mu_2 = \mu_0$, $\xi_{c2} = -4\xi_{c1}$, $L_2 = 2L_0$
   b. Determine the location of the center of the first stop band. Also, determine the bandwidth of this stop band.
   c. Same data as above but $\xi_{c1} = 0$, i.e., the layers are not chiral. Determine the first stop band and its width.

Index

A

Adiabatic analysis
  of modified source wave, 266–267
  for R wave, 147–150
  Airy functions, 29, 57
Amplitudes, 132, 259, 264, 267
Anisotrophy complexity, 107
Asymptotic frequency, 131
Attenuated total reflection, 38
Attenuation constant, 224
Attenuation length, 180, 185
  collision frequency and, 181
Attenuation of waves, in lossy medium, 204
A wave, 207
  incident wave frequency and, 208

B

Bessel-like functions, 74
Bloch wave condition, 30
Boundary value problem, 7
Bragg reflectors, 34
Brewster angle, 24, 25
Brewster mode, 37
B waves, 123, 207

C

Cerenkov wakes, 164
Chiral medium, 164–165
Coherent radiation, 231
Cold isotropic plasma, 1
  basic field equations for, 93–94
  time-varying and space-varying, 1
Cold plasma, 1
  lossless, 132
Collision frequency, 181, 183, 199
  attenuation length and, 182
  low-loss magnetoplasma of, refractive index for, 224
  time discontinuity and, 202
Controlled-fusion containment, 173
Current density, 169, 171
Current density vector, 261–262

D

Damping time constant, 69, 123, 125, 204
  defined, 68
  energy velocity and, 182
  gyrofrequency and, 177
  normalized incident wave frequency $vs.$, 70
  for plasma low-loss, 67
Decay time constant, 177, 183, 193
Dielectric
  defined, 8, 160
  dispersive, 160
  plasma as a, 8
Dielectric constant, 8–9, 15
  for extraordinary wave, 103
  of isotropic plasma, 11
  of Lorentz medium, 161
  for L wave, 98
  variations in, 14
Dielectric-electric spatial-boundary problem, 15
Dielectric tensor, 103, 105–106
Dispersion relation, 249–250
Dispersive dielectric, 160
Doppler frequency shift, 157

E

Eckersley approximation, 224–225, 227–228, 235–236, 239
E-formulation, 145
Electric field
  incident, 198
  magnetoplasma medium, 112
  of reflected waves, 88
  scalar, 81
  of source wave, 22
Electromagnetic radiation, time-switched magnetoplasma medium, 185
Electromagnetic signal, reflection of, 157
Electromagnetic wave, monochromatic plane, 245
Electromagnetic wave transformation, 94, 99
Electron density, 41
  rapid rise of, 173
  time $vs.$, 66

RETURN TO ➡ **PHYSICS LIBRARY**
351 LeConte Hall
642-3122

LOAN PERIOD 1	2	3
1-MONTH		
4	5	6

ALL BOOKS MAY BE RECALLED AFTER 7 DAYS
Overdue books are subject to replacement bills

DUE AS STAMPED BELOW

This book will be held in PHYSICS LIBRARY until OCT 12 1998		
AUG 2 0 2001		

FORM NO. DD 25

UNIVERSITY OF CALIFORNIA, BERKELEY
BERKELEY, CA 94720